Insight and Industry

Inside Technology

Wiebe E. Bijker, W. Bernard Carlson, and Trevor Pinch, editors

Insight and Industry: On the Dynamics of Technological Change in Medicine
Stuart S. Blume

Artificial Experts: Social Knowledge and Intelligent Machines
H. M. Collins

Viewing the Earth: The Social Construction of the Landsat Satellite System
Pamela E. Mack

Inventing Accuracy: A Historical Sociology of Nuclear Missile Guidance
Donald MacKenzie

Insight and Industry
On the Dynamics of
Technological Change in Medicine

Stuart S. Blume

The MIT Press
Cambridge, Massachusetts
London, England

©1992 Massachusetts Institute of Technology

All rights reserved. No part of this book may be reproduced in any form by any electronic or mechanical means (including photocopying, recording, or information storage and retrieval) without permission in writing from the publisher.

This book was set in New Baskerville by The MIT Press and printed and bound in the United States of America.

Library of Congress Cataloging-in-Publication Data

Blume, Stuart S., 1942–
    Insight and industry : on the dynamics of technological change in
    medicine / Stuart S. Blume.
        p.   cm. — (Inside technology)
    Includes bibliographical references and index.
    ISBN 0-262-02332-6
    1. Medical innovations.   I. Title.   II. Series.
R855.3.B64   1991
610' .28—dc20   91–18801
                                                                 CIP

To the memory of Marta Blume (Schwarz)

Contents

7

Image of Truth Newborn   225

Preface and Acknowledgments

This book has curious antecedents. In 1978, after having worked for a number of years in the sociology of science, in science policy, and, more recently, in social policy, I was becoming involved in what was for me a rather new area: health policy and politics. The Committee on Social Inequalities in Health, chaired by Sir Douglas Black (President of the Royal College of Physicians) had been set up in 1977 to look into the extent and sources of social disparities in health in Britain. As scientific secretary of this committee, collecting data, drafting analysis, and participating in sometimes heated discussions of the committee's recommendations, I came to share the concerns with both health and inequality that underpinned the longstanding dedication of its members. From Douglas Black, Cyril Smith, and, above all, Jerry Morris and Peter Townsend, I acquired a more informed, but also more questioning, interest in the relations between medicine and the health of the population. The committee's final deliberations, which included a divisive and painful debate on the desirability of funding essential improvement in living conditions through economies in medical expenditure, had a considerable impact on me. I subsequently reflected long and hard on how and why this occurred (Blume 1982). A little later, just as the final touches were being put to the report, the Labour government that had established the committee was defeated in a general election, and the incoming Conservative government attempted to bury the committee and all its works (see Townsend and Davidson 1982).

It was in this period that the idea of a study of technological change in medicine was given to me by Rudolf Klein, now Professor of Social Policy at the University of Bath. Knowing of my earlier interest in science and technology, and of my work with the Black Committee, Klein invited me to prepare something on "medical technology" for a conference he was helping to organize in Berlin. His suggestion came as something of a surprise. Although the Black Committee had discussed the costs of pharmaceuticals at length, I had never thought systematically about what lay behind their existence. Through what processes of technological change did new drugs, new devices, actually come into being? Why some and not others? Gradually, I realized that this topic would not only enable me to pick up some of my older interests but would also allow me to continue to think about social and political aspects of medicine. I took up Klein's suggestion and presented the first preliminary results of my (spare-time) work in Berlin in 1981. Subsequently, the Black Committee having reported, I looked into the possibilities of continuing with this research.

It was at that point that the Leverhulme Trust generously agreed to finance the first year and a half of my work, while the London School of Economics, with which I had been associated during the life of the Black Committee, agreed to provide me with further hospitality. Here the first of many debts that have been accumulated in the course of writing this book needs to be acknowledged. Without the support of Dr. (now Sir) Ronald Tress, then Director of the Leverhulme Trust, and of the Leverhulme Trustees, this book might never have been written. Professors Brian Abel Smith and Robert Pinker of the Department of Social Administration at the LSE provided encouragement, support, and access to an intellectual climate from which I profited greatly.

My work thus began within a context in which health policy figured large. The central questions that I wanted to address concerned the possibilities of harnessing the power of technology to the goals not so much of medical practice in any narrow sense as of health policy conceived as the outcome of a broader political process. I had learned that the two did not necessarily coincide. Did new medical devices always have to be more expensive, more specialized, more complex and difficult to run than old ones? Couldn't we have an innovation process in which things got simpler, hospital care became more instead of less accessible, and money was saved for nonhospital care? It soon became clear that posing questions like these necessitated empirical analysis of the forces actually bearing on the development of new medical technology. Stocking and Morrison, in their study of the CT scanner, had come to similar conclusions (Stocking and Morrison 1978). Although I rapidly determined upon a study of the development of a number of diagnostic imaging technologies, it was by no means clear what the dimensions of the analysis had to be. The necessary intellectual tools were not immediately accessible. There was a clear relationship with issues raised in the very disparate areas of medical sociology on the one hand and the economics of innovation on the other, but the concepts that I would need to structure the analysis remained elusive.

Moving from London to Amsterdam in 1982 influenced the course of the work in many ways. My new duties left far less time for my own research. More positively, the concepts I needed now started to become clear. The change in intellectual focus, discussions surrounding the applicability to technology of notions becoming popular in the sociology of science, helped greatly in that respect. But the newly clarified conceptual structure was also coming to dominate. I realized that the focus of the book was changing, and I was increasingly seeking to use the analysis of technology in medicine as a way of posing more general questions about the steering of technology. Perhaps this was inevitable, though it made the writing difficult, not only in the usual ways, but emotionally too. I hope that something of my original concerns and commitments remains so that the book retains relevance for debates regarding health and medicine. I say this not only out of a sense of loyalty or debts to be repaid; it has also to do with my personal intellectual and ideological concerns regarding the field in which I now principally work. To whom should studies of technology be addressed? I think that there is much to be said for the view that work in the social study of science and technology must be exposed to rigorous and extensive critical debate, not least within a more general social scientific domain, if the field is to avoid becoming either an esoteric cult or a branch of policy analysis.

The past eight years have been stimulating, frustrating, challenging, and despairing in turn. They have never been dull, and I want to thank all my colleagues and students in the Department of Science Dynamics for making it perhaps the most pleasurable of places in which to study the sociology of science and technology. Anja Hiddinga, who has witnessed my responses to the challenges of these years and who has ministered to my attempts to come to terms with an unfamiliar academic culture, deserves very special thanks.

Among those who have read various versions of the manuscript, in whole or in part, and from whose comments, criticisms, and counsels I have particularly profited are Olga Amsterdamska, Tom Burns, Annetine Gelijns, Eda Kranakis, Terry Shinn, Wesley Shrum, and Rosemary Stevens. I would like to thank them all, adding, as is usual but in this case is certainly appropriate, that all faults of inquiry, methods, and reasoning are mine alone. Thanks are also due to Paulette Röring, Tini Tonneman, Olga van der Veen, Miep Wuck, C. Holms of Philips Medical Systems, and the photographic services of the University of Amsterdam, all of whom provided invaluable help with production of the final manuscript. It has been both pleasurable and rewarding to work with the editors of this series, particularly Wiebe Bijker with whom my contacts have been closest.

I have extracted the titles of the first two chapters from Book XIII of William Wordsworth's remarkable autobiographical poem, *The Prelude,* which he wrote in 1805 and then rewrote in 1850. I would like to give the appropriate phrases in context:

. . . Imagination, which, in truth,
Is but another name for absolute power
And clearest insight, amplitude of mind
And reason in her most exalted mood.

Calling upon the more instructed mind
To link their images with subtle skill
Sometimes, and by elaborate research
With forms and definite appearances
Of human life

Insight and Industry

1

Absolute Power and Clearest Insight

The Dilemmas of Medical Progress

A 1987 article in the *National Geographic* magazine vividly describes some of the wonderful new devices with which doctors today can "watch vital organs at work, identify blockages and growths, and even detect warning signs of disease." These devices are "changing the face of medicine." To find out what is really going on, what these machines are really capable of, the author has taken his cameras into the hospital. "And now," he reports, "I was subjecting my beyond-middle-age body to the scrutiny of one of the marvelous machines—never dreaming my reporting would suddenly become deadly serious."

Preliminary tests have suggested that the journalist, Howard Sochurek, might indeed have something wrong with his heart. He is about to be tested on one of medical science's most recent wonders, a Positron Emission Tomography (PET) scanner:

I lay down on a table and was slowly moved into a huge two-ton metallic doughnut with a center hole just large enough for me to squeeze through—the 1.6 million dollar PET scanner. . . . Fifteen minutes of scan time took place while the heart was beating normally. I heard cooling fans humming in the scanner and the beep beep beep of the EKG machine, which monitored me continuously. My head throbbed with anxiety, although Mary and Dr. Gould hovered round me.

Consider what is involved in these extracts. First, there is wonder: It is with a sense of the miraculousness of it all, he makes clear, that he set out to write his story. But simultaneously there is a second emotion: fear. Being put in the machine with the hint of something wrong, despite the attentive nurse and the six-foot-six Texas doctor, makes the writer's head throb with anxiety. The two responses seem perfectly compatible.

How does such a machine succeed in arousing such powerful emotions? This is a cultural question that can quite properly be asked of all sorts of technologies: A great medical machine is by no means unique in producing a mixture of wonder and fear. Much modern technology, and perhaps technology as such, is awesome in just this way. From the nineteenth century onward many writers and artists have sought to convey this notion, Mary Shelley's *Frankenstein* being one oft-cited example. Related concerns have been expressed by philosophers such as Martin Heidegger and Hannah Arendt, who have argued the need theoretically and practically to confront "technology as such," suggesting that this, rather than any specific technology, threatens the history both of Man and of Nature (Rapp 1981). It has also been pointed out that Western culture, and particularly North American culture, shows many signs of a fundamental adaptation to technology that it began to make in the nineteenth century (Marx 1979). Not all thinkers take an equally sanguine view of the consequences of this adaptation. Yet to understand the place of technology in modern medicine, of a technology that seems to have become almost ineffable, it is not enough to take refuge in a general cultural-philosophical analysis of technology, important though this is.

Certainly part of the task of understanding has to be to strip technology of its ineffability. One of my objectives will be to show that even the most awe-inspiring artefact is just that: a thing made by human beings driven or inspired by certain goals, desires, or aspirations. We shall be concerned with why it is that certain sorts of medical devices emerge and become embodied in medical practice. There is nothing inevitable in the design, the form, that a technological vision—whether of a steamboat, an undersea ship, flight, or communication over long distances—comes to take (Hindle 1981). In this book we shall be concerned with the processes of social shaping that give form to visions of this sort. And the visions we shall focus on are ones literally of insight: new ways of looking into the human body.

In Sochurek's article, optimism triumphs. It is an optimism rooted in a willing acceptance of the inevitability and at the same time of the benefits of technological advance. The article ends, "How often while preparing this story did I hear: 'I couldn't have saved this patient ten years ago.' Because of the computer and the new tools of machine vision, many more lives will be saved in the years to come, and the quality of all our lives will be improved."

Our beliefs about medical technology cannot adequately be deduced from our beliefs about technology in general because they incorporate our particular cultural views about medicine. Medicine saves lives. Perhaps one day my life or your life will depend upon new drugs, devices, or operations. This creates a distinction from our attitudes toward personal computers or supersonic transport or the new possibilities of home entertainment. Because we see medical technology as serving the benign purposes of medicine, our attitude toward it is special, hallowed. At the same time the very marvelousness of the technology provides reassurance. It has been shown that British women would rather have themselves screened for breast cancer by x-ray than learn techniques of self-examination (Calman 1984).

Most of us welcome both the possibility of handing over responsibility for our health and the reassurance that technological medicine seems to provide. Yet, thinking about the future, many share a certain discomfort regarding the apparently uncontrolled "technologization" of the health care system. With time, fewer and fewer people will be in a position to regret the passing of the family doctor—the friend of the family, always willing to pay a reassuring visit. Yet the fact is that much of the reassurance that people sought from their family doctors concerned coping with sickness and had little to do with the clinical entities of advanced medical science. Today only the hospital specialist, the master of advanced technology, seems to have the authority to reassure, yet coping with sickness is not a matter with which the specialist is typically concerned. Something has, perhaps, been lost.

Some see that something as medicine's humanity. Indeed some sociologists and particularly feminists argue that technological change is important precisely by virtue of its reinforcement of existing power relations (Ehrenreich and English 1973, Navarro 1986, Zola 1975). Related to this argument is the ethical critique of modern medicine, manifest most clearly today in discussion of the possible uses of genetic screening. There is something fundamental at stake here. As one critic has put it, "the basic issue is to keep taking ethical readings on our research approaches so as to be sure that we do not destroy the moral fabric of society in our zeal to improve its physical fabric. Some scientific gains may only be available at a price we are unwilling to pay" (Lasagna 1972, 274).

A third area of concern is financial. To medical policy makers who must deal with the cost of new technology, new wonder drugs, and complex new procedures such as organ transplants, each of these advances carries problems as well. One frequently suggested means of dealing with innovative but costly techniques and equipment has been rationing. After all, two contributors to the high cost of medical care have been shown to be the unnecessary duplication of expensive facilities and the overuse of those facilities. Shouldn't it be enough to allow everyone "reasonable access" to advanced facilities? A number of countries have in fact tried rationing systems in the attempt to rationalize, or limit, the diffusion of new instruments and procedures. In the United States, for example, a number of legislative developments over the past twenty years have addressed the adequacy of use of procedures, distribution, and the effectiveness and safety of devices. Yet this "political rationality" seems frequently to succumb to other interests, perhaps to another rationality.

Part of the trouble has been that the nature of the innovation process in medicine has changed. Health economist M. E. Klarman notes that "The late 1940s and early 1950s were marked by the introduction and widespread diffusion of many new drugs, particularly the antibiotics, which had a pronounced effect on the length and severity of infectious diseases." However, "Since the mid 1950s, advances in medical technology have not brought about a similar improvement in the ability of physicians to improve health. Renal dialysis, cancer chemotherapy, and open heart surgery may achieve dramatic effects in particular cases, but bring about only marginal improvement in general indexes of health. Moreover, the early advances tended to be physician-saving, while the later ones were characteristically physician-using" (Klarman 1974).

But more significant is that other rationality, which categorizes the way most physicians view medical care, and which shapes the way physicians typically deploy advancing technology. According to this view, rationing access to the latest technology is an unacceptable denial to some patients of the potential benefits of medical progress. We shall see that hospitals too may have institutional interests in securing possession of the most advanced devices and in being able to offer the newest procedures. This view resonates with the beliefs held by much of the population. There is evidence from American studies that, just as patients want to feel able to rely on advanced technology, so too they tend to identify use of technology with quality of care. Wagner and Zubkoff have noted "patient demands for technological sophistication as a surrogate for high 'quality' care" (1982, 167). But while, by and large, high-technology medicine is identified with high-quality medicine, some people see negative consequences in this dependence on technology. Feminist criticism of the routine use of ultrasound in obstetric care is one example. Even some physicians have criticized their growing dependency on batteries of increasingly specialized techniques. And there are indeed a number of clinical studies demonstrating how techniques such as fetal monitoring might be more effectively deployed through more selective use (see, e.g., Neutra et al. 1978).

An eminent professor of neurosurgery, B. Jennett, writes that "High technology . . . appears to be relevant to few of the major problems clamouring for attention in medicine today; even for those patients who do become recipients of high technology there is anxiety that it may sometimes do more harm than good" (Jennett 1986, 2). As a neurosurgeon, Jennett is, he confesses, "an archetype of high-technology man." But he is well aware of the many reported failings in the introduction of and reliance on increasingly complex techniques, and of the sometimes dubious evidence for their benefits to health care. Like many sympathetic critics, he seeks a measured and rational middle way in the use of technology.

The case studies in this book show that the reaction to new medical technologies is anything but rational and measured. But to understand these reactions, we need to understand the overall factors that shape the response of the medical community to technological innovation. After exploring these overarching elements of technologization, this chapter will conclude with an introduction to the first imaging system that had a strong effect on medical practice: the use of x-rays.

In What Ways Is Medical Practice Dependent on High Technology?

When a new technology (a device or drug) becomes commercially available, and looks promising, it will diffuse and be gradually absorbed into practice.[1] Studying both its diffusion and its eventual impact on medical practice can teach us about the dependency of medical practice on technology. Studies of diffusion processes cast light on the sorts of evidence on which investment decisions are based, thereby hinting at the interests and values that lie behind these decisions: profit, where applicable, to be sure, but also something else. What they show is that the interests, commitments, and beliefs of physicians and the institutional interests of the hospitals in which they work (which are by no means a simple summation of physician interests) are linked in a number of ways to the purchase and introduction of new technology.

There have been a number of quantitative studies of the diffusion of new medical technologies. Almost all produce a characteristic S curve for the *pattern* of diffusion, but there are major differences in the *rates* of diffusion (see figure 1.1). If we could compare technologies, focusing on the characteristics that affect their rates of diffusion, then we would be in a position to say more about the nature of the dependence of medicine on technology. The CT scanner has often been cited as an example of especially rapid diffusion. What factors are said to have been behind its remarkable rate of diffusion? The first such scanner, for use on the head, was apparently safer than the rather risky and unpleasant array of previously used techniques, such as pneumoencephalography, which involves injecting air into the brain. But what about the "whole-body" version, which diffused equally rapidly without these obvious indications? That scanner was certainly profitable, and it was also *compatible* with radiology practice, in the sense that it seems not to have required major investment in retraining. These factors all seem reasonable but hardly the basis for a broad explanation of the sort that might be based on comparisons of a variety of technologies.

More generally, Banta has described a variety of factors that seem to hasten adoption: relative advantage over previous methods; compatibility with the adopters' values; the possibility of testing the innovation on a limited basis; the complexity of understanding and using the innovation; and the visibility or demonstrability of the results to others (Banta 1984). These notions are heuristically useful but by no means explanatory. What, for example, are those adopters' values with which an innovation must be compatible?

Economic interests play an important role in the diffusion of new technology, and economics is central to any discussion of the diffusion process. When resources are limited, it is likely to be the cost of one form of service that determines the resources available for other forms. Economic studies show both that high technology medicine is expensive and that technology has an important relationship with the cost structure of medicine. After all, unlike much

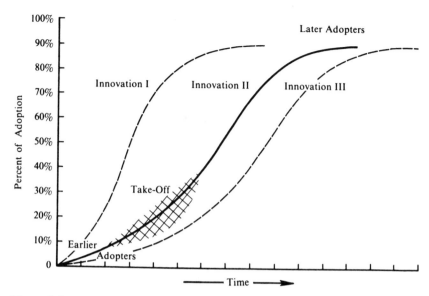

Figure 1.1
The diffusion process: Characteristic curves showing diffusion of three innovations. Source: Rogers 1983, p. 11. Courtesy of Macmillan Publishing Company, N. Y.

other innovation in "process plant," innovations in the high-technology armamentarium of medicine are rarely (if ever) designed with a priority objective of cutting the cost of providing medical care (Iglehart 1977). Cost savings may be a consequence of introducing a new technology into a particular hospital (for example, through reduced demand for older procedures), but that is rarely the objective of introducing the technology, still less a priority in its development.

Precisely because of what is at issue, and because of the authority of economic argument in policy discussion, the economics of technical change in medicine are controversial.[2] There is certainly disagreement regarding the contribution of new technology to the rising costs of medical care. "What generates most controversy," according to Jennett, "is the relative cost of different types of technology and comparison with activities that are less dependent on technology." Thus, if you compare the costs of a typical hospital in-patient day in different sorts of wards, you find, for example, that a typical day in geriatrics or rheumatology is relatively cheap, one in obstetrics and gynecology about average, and one in general surgery expensive (Jennett 1986, 155). But if you also take account of how long patients stay in a hospital, the picture changes. In the hospital described by Jennett, obstetric patients averaged stays of 5.5 days, general surgery patients 7.6 days, and geriatric patients 26 days. So obviously the total costs per patient are quite different from the costs per day. "Some specialities with high unit costs . . . make little impact on overall expenditure and the same goes for some high technologies (transplantation of kidney, heart or bone marrow) because relatively few patients are involved."

The general thrust of Jennett's account is to suggest that salaries are a much more important factor in the rising cost of hospital care than is technology. A number of economists have expressed a similar opinion, though there is certainly no consensus.

If we look specifically at the economic *incentives* underlying the introduction of advanced technology, however, the picture is clearer. In the United States in particular, advanced technology is extremely *profitable.* There is therefore an economic interest in making use of sophisticated new procedures and devices. "The financial incentive occurs most strikingly in private practice settings in which physicians own major pieces of equipment, such as x-ray and electrocardiogram machines, and have a financial stake in promoting use of this equipment" (Myers and Schroeder 1982). As an example of the profit potential of expensive devices, consider the following findings from a study of the policy implications of the CT scanner carried out in the United States at the end of the 1970s. Typical costs to a hospital of a CT brain scan—taking account of capital cost (spread over a few years), of materials, and of salaries—ranged from $79 to $173. Typical charges ranged from $240 to $260. Overall average charges for CT (head) examinations exceeded expenses by 39 to 229 percent (see OTA 1978, 88–89).[3]

Of course, much more is at stake than financial interests. Recourse to new technology is *normatively sanctioned,* that is, sustained by a complex set of values and interests. It expresses an important commonality between the values of hospital physicians and those of the institutions in which they work. Consider the following account by a professor of radiology from Harvard Medical School:

The need to obtain increasingly sophisticated anatomic and physiological information for good patient care sometimes can make us desperate to encourage new technology. Just as many patients will grasp for any therapeutic straw . . . so we in clinical neuroscience might initially endorse a new technology only on its promise. For example, two or three years ago an enthusiasm for positron imaging began to swell in neuroscience circles.

Without critical inquiry neuroscientists allowed themselves to believe that positron imaging could do more than it was capable of doing. Now two or three years down the road, the impatient neuroscience community wants to know why positron imaging has not produced more. . . . The fact is that we are only now assembling the forces, instrumentation and strategies to really explore what positron imaging can do. (Ackerman 1981)

The argument here is that the exigencies of patient care—the need to take action—encourage unrealistic expectations, which are then bolstered by claims well beyond what could be supported from the scientific literature. Sociological studies bear out Ackerman's view that significantly more than profit is involved in the enthusiasm with which physicians frequently greet new technologies. Yet to understand the dynamics of the process we must also look at how these views are reflected in hospital behavior. Largely speaking, hospitals try to provide the facilities that physicians want: again not only as a result of economic calculus.

Physicians are highly influential in hospital decision making, so that their enthusiasm is of some importance. There are now a number of accounts suggesting that a medical mode of control has dominated the American hospital in recent years. The administrative hierarchy, which might seem to have an interest in cost-containment measures opposed to physician interests, largely speaking does not (Joskow 1981, 31–36). Given the reimbursement systems in operation through the 1960s and 1970s, argues Joskow, administrators could reasonably assume that costs would be recouped. Hospital organizations have either "weak administrators or administrators who consider it their job to build a bigger, more sophisticated, and more prestigious hospital. The administrator interested primarily in organizational changes designed to reduce costs is not likely to survive long in this system" (p. 34). Work by Greer (1984) throws more light on physician involvement in decision making. Looking in detail at decisions taken by hospitals to purchase new devices, Greer comes to dismiss two common notions as too simplistic. It is not the case that physicians urge these purchases either out of their own financial interest or because their training has caused them to prefer high-technology medicine. What Greer does find, perhaps not so surprisingly, is that there are quite distinct and distinguishable specialist interests associated with the acquisition of new technology. The main supporters of investment in new technology in American hospitals tend to be members of particular hospital-based specialties, including pathology and radiology.

What kinds of considerations are involved in decisions to purchase new instrumentation? Greer found that the kind of decision involved depended on the kind of facility in question. At one extreme were small pieces of equipment "desired by individual physicians for use in their own clinical practice." Here decision making was dominated by medical issues and priorities and largely restricted to the medical hierarchy. At the other extreme were big pieces of equipment like CT scanners or—of still greater scope in its implications—decisions to introduce an intensive care unit or begin open heart surgery. Here administrative and financial officers and even boards of trustees might well be involved. For what is at stake here is something other than quality of care: it is a combination of "market share," staffing policies, "organizational image," and considerations of "institutional traditions or ethnic pride."

A study by Russell of the diffusion of seven technologies among a large sample of American hospitals provides some basis for looking at the characteristics of technologies and adopters simultaneously (Russell 1979). The comparative dimension proves highly instructive. As might be expected, Russell found that, in general, "large hospitals adopted a given technology sooner, on average, than small hospitals." Particularly interesting was a comparison based on involvement in physician training programs (residency programs), medical school affiliation, and involvement in research. All of these were associated with more rapid adoption of new technologies. But still more significant was the fact that the effect of these variables proved to depend upon the technology in question. Among the technologies in Russell's sample, the research and teaching hospitals were particularly inclined to adopt intensive care and diagnostic radioisotopes at a rapid pace. Russell explains this in terms of the relative interest for research of these techniques: "This outcome is consistent with the nature of the technologies, both of which are widely used, but at the same time scientifically prestigious. Diagnostic radioisotopes were adopted faster the higher the level of research grants to hospitals, while the number of intensive care beds was measurably affected only by very high levels of research spending" (Russell 1979, 159).

The introduction of (advanced) new technology into hospitals thus reflects the influence and beliefs of important segments of the medical community. Yet there are also distinct organizational interests with which these medical views are connected. The early introduction of new devices or procedures may be associated with commitment to a particular form of care, or to a reasoned assessment of the interests of patients and of hospital. Gradually, however, adoption by more and more hospitals may come to be based less on rational assessment of the task at hand than on a kind of copying behavior. Such a process of homogenization, or "institutional isomorphism," has been suggested to be a general feature of organizational behavior (DiMaggio and Powell 1983). Gradually, according to this view, adoption of innovations ceases to be a means of improving performance and becomes a means of achieving legitimacy within the "organizational field." Professionals, mobile within such fields, are an important element in this process of growing isomorphism. Indeed, DiMaggio and Powell claim that the more professionalized the labor force of a class of organizations, the more competition between them will take place on the basis of status claims rather than efficiency of resource use (p. 154). In the case of hospitals this seems to be so. According to Joskow, "In most areas of the country several hospitals serve overlapping patient populations. If physicians cannot get what they want from one hospital, they will try to get it from another. A single administrator oriented towards cost control and wider considerations of social efficiency may find his efforts rewarded with the departure of top physicians and their patients. . . . For the hospital to survive and grow, the administrator must endeavour to meet or beat the competition. He cannot do it by providing care at lower cost so he must do it by providing the services that can attract physicians and their patients" (Joskow 1981, 35).

Advanced technology is vital for modern medical practice not only because for many physicians it is associated with, or seen as essential to, the highest quality care. Nor is it only because technology is, or can be, very profitable. No less important are the sorts of factors teased out in Greer's and Russell's studies and captured more abstractly by DiMaggio and Powell. Technology is a source of status, of distinction—it is a means for physicians to do and publish research, thereby showing not only the quality and importance of their work but also (on occasion) their privileged access to the most modern and desired equipment (see Hillman 1988). For hospitals, new technology serves to attract research grants and the best clinical scientists, and to bolster their images as centers of the highest standing. Over the years the very structure, the social order, of the hospital has come to be shaped by the technologies in use. At one level are specialists such as pathologists and radiologists, whose status depends on the gleaming and powerful instruments they control. Lower in the hierarchy are those who depend for their livelihood—and position in the social order—upon their intimate familiarity with the equipment they use, as in the case of laboratory chemists and radiographers (Larkin 1983, Cockburn 1985). Even so apparently humble a matter as servicing medical equipment, keeping it running, has often been the source of bitter disputes between skilled technical groups in the hospital. Indeed the very origins of hospital physics are intimately bound up with the need for expert supervision of the use of radionucleotides in the hospital.

The structural dependency of medicine on technological advance—and I refer here specifically to devices and drugs—cannot be understood, however, in terms of the social and cultural order of the medical care system alone. It is bolstered by the fact that commitment to technological progress is shared with the producers of new technology, that is, with industry. In the post–World War II period medical equipment has become very big business, and today physicians, engineers, manufacturers, and patients make common cause in supporting technological advance in medicine (Eden 1984, 53). Many physicians perceive the significance of this community of interest. For example, a University of Washington radiologist, responding to a critique of the limited and "descriptive" data on which the diffusion of magnetic resonance imaging was supposed to have been based (see chapter 6), wrote, "But was the dissemination of MR technology based solely on such data or were other factors involved as well? One notes that corporations invested billions of dollars in development of MR equipment and that a large number of financial administrators invested millions of dollars each in MR scanners. One notes also that a rather tough and unforgiving marketplace has successfully borne this level of investment in MR" (Shuman 1988). In DiMaggio and Powell's terms, the "organizational field" encompasses more than the hospitals (and their patient populations) alone. Indeed they define it to include "those organizations that, in the aggregate, constitute a recognized area of institutional life: key suppliers, resource and product con-sumers, regulatory agencies, and other organizations that produce similar services or products" (DiMaggio and Powell 1983, 148). Innovation in medical devices could not have the character it does, whatever the values and commitments of the medical profession, if this were not so. That the *common* interests of segments of the medical care system and segments of manufacturing industry are crucial to technological innovation in this field is a notion central to the argument of this book.

Historical Roots of the Dependency of Medicine on Technology

In the second half of the nineteenth century medicine underwent a major transformation—in fact a succession of intellectual and practical transformations wrought by people of genius and convic-tion and sustained by broad economic and social shifts in the societies in which they lived and worked. The details of what was achieved by men such as Bernard, Virchow, and Koch is beyond the scope of this book. But though there were many differences among them in the precise scientific basis they wished to give to medicine, all were concerned to found medical practice in rigorous scientific theorizing and investigation, akin to those cornerstones of the physical sciences. Bernard, for example, sought not only to root out all traces of old vitalist "superstition" but also to establish a physi-ological basis and a scientific method for medicine (see Figlio 1977, Geison 1979).[4] His determination to (re)found medicine in scien-tific physiology led him to oppose no less those who espoused the epidemiological/public health perspective.

What Bernard, Virchow, Koch, and Pasteur did was to move the essential locus of medicine into the laboratory. Here would be made the discoveries—by physiologists, by bacteriologists, and by biochemists—on which medical progress depended. But not all in medicine who sought its modernization looked on this shift in locus with equanimity. Among those who viewed the victories of the laboratory doctors with some concern were a number who, while recognizing the power of the scientific method, began to wonder about the possibility of bringing it into the hospital ward. Might it not be possible to be scientific, and to reap the profits of being scientific, without distancing medicine from its natural focus on the human beings under care? From these deliberations there emerged, first in Germany, a notion of clinical research. With this notion, hospital medicine could be secured against a laboratory science that threatened to dominate it. In 1880 there appeared the first issue of a new journal *Zeitschrift für Klinische Medizin*: "a very manifesto of clinical science, the journal proclaimed the autonomy of clinical medicine as a discipline" (Maulitz 1979). Clinical science would have its own, rigorously scientific methods of investigation. Instead of "discerning the morphologic lesion or the isolated strain of pathogenic microbe," clinical scientists would investigate "those functional parameters of the human body which, in health or disease, were uniquely susceptible to study on the hospital ward." No less scientific, then, they would turn their investigative powers to the patient in the ward, rather than to an abstraction of sickness taken away to the laboratory.

The relevance of this nineteenth-century revolution for understanding the relations between modern medical practice and technology is twofold. In the first place we shall see that clinical science brought with it, eventually, a transformed place for technology. Precise and reproducible measurement, precise and reproducible observation, could not be accomplished adequately with the unaided senses of the physician. The establishment of a corpus of clinical scientific knowledge required a technology of observation and measurement. Second, the emerging clinical science was successful because it came to be associated with a transformation in the institutions of medicine: the hospitals, the professional reward system, and education. Leading institutions, to remain so, had to espouse the new medical ideology. Naturally, institutional transformation was not everywhere the same, and there were differences between societies in how exactly the old medicine gave way to the new scientific practice (Hiddinga 1987). But in general "scientific medicine" provided a rallying cry for reformers. According to Rosenberg, "The most "advanced" and influential reformers of the 1870s and 1880s saw the principal function of the new model hospital as that of providing a necessary home for scientific medicine.... In the 1890s, it had already become apparent that scientific reputation would serve as an advantage in attracting the brightest and most ambitious young medical men and, it soon became clear, the support of individual and corporate benevolence" (Rosenberg 1981, 48–49).

The creation of an ideologically and practically distinctive medical practice went hand in hand with the creation and diffusion of a new medical technology. Instruments were being created that not only served in various senses to "distance" the physician from the patient but also provided the physician with (apparently) more precise, quantifiable, and recordable information on the patient's condition—a clinical armamentarium to parallel that being developed in the laboratory by the bacteriologists. The development of the medical thermometer is a well-known example. Pioneered by Hermann Boerhaave in the 1700s, the medical thermometer was used by Boerhaave and his pupil de Haen in substantial studies of the thermometric correlates of disease (see Reiser 1978, 112–120). But only with the appearance of Wunderlich's great work *On the Temperature in Diseases*, in 1868, was the basis established for a serious role for the thermometer in diagnosis. Wunderlich presented thermometric data on almost 25,000 patients, as well as instructions for the medical use of the thermometer and a discussion of temperature variation in thirty-two common disorders. He was an enthusiast, arguing that thermometry was superior to all other diagnostic techniques. But the superiority was inherent not only in the significance of temperature in health and disease: it was also to do with the fact that the thermometer permitted *delegation* of the practical part of the act of diagnosis. An assistant could do the actual measurement, leaving the physician to apply diagnostic skills to the interpretation of the data generated. In other words, although the medical thermometer had existed for over a century, it was only at the end of the 1860s that seemingly persuasive data and indications for its use appeared. Not everyone was persuaded, though, certainly not at first.[5] Significantly, it was in this same period that other clinical instruments were beginning to make their appearance, including the ophthalmoscope (developed by Helmholtz), the sphygmograph (developed by Vierodt and by Marey for measuring the pulse), and the sphygmomanometer for measuring blood pressure (announced by von Basch in 1884).

These developments, and the transformations in values, standards, and practices that lay behind them, did not take place wholly without resistance. The early history of some of the new institutions shows that not everyone shared a common view of what the new scientific medicine was to be (Harvey 1981, chapter 4). Christopher Lawrence's account of this resistance among leading English medical circles throws light on what is undoubtedly a crucial feature of the relations between medicine and advanced technology even today.

For the elite of English medicine, Lawrence argues, science provided a vocabulary, a rhetoric, with which to bolster professional authority (Lawrence 1985). But the extent to which practice was actually to be made dependent on science was entirely another matter. The status of the medical elite in London in the second half of the nineteenth century was based to a considerable extent on broad education, mastery of the classics, the social skills of the gentleman—in short, on character and breeding. That medical practice should become the application of the results of science, that it should be governed by precise rules for the application of these results, was both a threat and an affront to the leading figures of London (and English) medicine. There were two sorts of attempts to use science to change medical practice Lawrence argues. In the first place there were those who argued that what was essential to medicine was its reasoning processes, and that for these to be equivalent to the reasoning processes of science meant that they should be acquired through an education rooted in basic science. But "such apparent reductionism was anathema to gentleman physicians, who defined clinical medicine as based on science but described its actual practice as an art which necessitated that its practitioners be the most cultured of men and the most experienced reflectors on the human condition."

There were others who laid less stress on reasoning processes and scientific education and more on the use of the new tools offered by science in the investigation of disease. It was the reformers, in the first instance, who made use of the tools that had been developed; they made little impact on the majority of doctors. Those who took them up, according to Lawrence, did so either because their real interests were in experimental physiology or because in one way or another they were trying to bring about some kind of institutional change in the organization and values of medicine. Thus while both use and advocacy of the new instruments of medical diagnosis were associated with the reform of medicine, the association was a complex one. In trying to make sense of the place of the newly emerging medical technology in the debate over medical reform, Lawrence ascribed it a legitimatory function: "Perhaps instruments were advocated not because they were valuable clinical tools, but because they were the children of the experimental sciences. If the advocates of those sciences could somehow show that the instruments had clinical application . . . " (Lawrence 1985, 517).

Historians of medicine increasingly incline to the view that the cultural more than the technical power of science provided the reformers with their crucial resource (Warner 1985). Consonant with this view, Lawrence's point is, of course, that the significance of (diagnostic) technology was in part that it stood for science-in-medicine and that the relation between the two was as much symbolic or metaphoric as one of descent. To demonstrate that they worked in the clinic was a way of legitimating the science on which they were based. In an elegant and insightful review of the historiography of relations among science, medicine, and society, John Harley Warner emphasizes the multiple meanings of science for turn-of-the-century medicine (Warner 1985). This perception is just as helpful in trying to make sense of present-day relations between medicine and science-based technology. Today, too, medical technology has many meanings—a crucial point for, as we shall see, it underpins the coalitions of interest though which the development of new technology is achieved.

The success of the reformers—the fact that diagnostic technologies did become part of the standard routines of medical practice, that medical education did become increasingly founded in the basic sciences, that research hospitals did become the cathedrals of medicine—still has to be explained. It is not that there was any self-evident superiority. More precisely, in the notion of medicine-as-art there was a ready vocabulary with which to dismiss claims to greater efficacy and promise, as Lawrence suggested. Part of the explanation of the eventual transformation of the practices and the institutions of medicine, and the new place for medical technology that came with it, has to be sought not in the details of clinical procedure but in broader social changes and in the new social policies that governments evolved in response. The new relations between medicine and technology were being forged in this greater fire.

There is a good deal of evidence for a continuing process of specialization in Western medicine, though it has naturally taken different forms in different countries.[6] Typically, the process is a complex one in which the emergent specialist groups of physicians must fight on a number of fronts for their authority. Starr has described the action of the specialists on two fronts as follows: "When specialists claimed that various techniques and procedures required their skills, general practitioners often found themselves damned in the same breath as non-physicians. The obstetricians who argued that midwives were inadequately prepared to handle deliveries frequently said the same of GPs. Hence two different conflicts were often taking place on the same terrain. The specialists sought to achieve ascendency over the non-physician specialists in their area—obstetricians over midwives, ophthalmologists over optometrists, anesthesiologists over nurse anesthetists, and so on. And they sought also to impress on the general practitioner the limit of his abilities" (Starr 1982, 223).

The development of medical technology has frequently been cited as a factor in the struggles for power that underlie processes of specialization, as in the case of the thermometer permitting physicians to delegate a routine part of their task. Looked at the other way round, we can also see how those whose expertise consisted precisely in a technical operation were made subordinate. Starr notes that the outcomes of the two struggles he discussed were in fact different. As of 1930, the general practitioners were able to protect their position, and the specialists' claims to all kind of privileges were successfully opposed. The nonphysicians, however, became "subordinated to the doctors' authority, usually permitted neither to practice independently of the doctor nor to interpret the results of tests or x-rays directly to patients." One can argue that, on the one hand, technology fosters specialization (further division of labor) in medicine, and that on the other hand financial and status interests of competing specialized groups foster further recourse both real and symbolic to technology—what one recent commentator has termed a "narrow professionalism" (Redisch 1982, 151).

A second contributor to the relation between technology and medicine can be located in the changing nature of what people expect from the medical care system. Stevens has summarized the factors involved in this change as follows: "The increase in public provision of health services was clearly one such factor. School health services . . . [the] experience of organized medical care in military service during World War I and the development of industrial health services had brought organized medical care, though of limited application, into the experience of millions of people. This experience was strengthened by increasing government participation through veterans programs. . . . These interlocking activities could be seen in the early 1930s as a movement crystallizing public opinion in the direction of regarding health as a social right" (Stevens 1971, 178). This change was coupled with a transformation in the nature of medicine. "While current social philosophies were being applied to health services, medicine itself continued to change. By the late 1930s, medicine had firmly established its status as a science. . . . Hormones, insulin, and vitamins, all twentieth century discoveries, were being applied in day to day living. Blood transfusion had become one of the commonest hospital procedures, together with a variety of x-ray procedures, electrocardiographs, basal metabolism techniques, and other diagnostic and treatment instruments which were transforming medicine from a complex mystery to an increasingly exact skill" (Stevens 1971, 179).

The connection between these processes is no accident of history. Social change, the demands of previously excluded groups for medical care, as well as progress were providing new challenges for medicine.[7] It was partly that the pattern of disease was changing; as old scourges were brought under control, others—allergies, diabetes, arthritis, diseases of the peripheral blood vessels—took their place. The significance of a given pathology for medicine is rarely a direct function of its prevalence in the population. Social factors—for example, who it is that suffers—intervene. Stevens's point is that people's concern with illness and pain was also changing. Medicine was faced with challenges that were both intellectual and matters of scale. Again science seemed the best approach to the one, technology to the other.

A third historical factor to play a significant role in the growing technological dependency of medical practice was the rise of life insurance. Davis (1981) has described the role of the expanding life insurance industry in demanding systematic and precise diagnostic information on a large scale. Doctors acting as medical examiners for life insurance companies were being confronted with patients who did not feel ill and who had not sought the physician out in the traditional way. This extension of the physician's role required that signs of disease be detected before they surfaced. For this, Davis writes, "physicians found it expedient to employ sensitive and precise methods of physical examination. This need or desire for the physician to evaluate an individual's bodily functions through physical diagnosis was an innovative diagnostic practice which emerged from changing social factors and available medical technology. The reasons for obtaining a physical examination in the nineteenth and twentieth century were, in addition to responding to a patient's feeling of discomfort, to obtain life insurance, to attend school, to enlist in the military forces, to obtain a job, to become eligible for a pension" (Davis 1981, 405). The new industrial social order, with extended public and private social provision for huge numbers of people, required information on the health status of much of the population. As the medical profession extended its roles and responsibilities to meet this need, both the nature and the scale of what was required led to growing reliance on technological means of diagnosis.

Ultimately, the interdependence of a medical practice increasingly dependent on and disciplined by technology on the one hand, and of a manufacturing industry seemingly being shaped by similar forces on the other, is a matter open to various explanations. When Simon Schaffer (1988) writes of mid-nineteenth-century astronomical observatories that "discipline and hierarchy inside . . . went hand in hand with the formation of a disciplined world outside their walls," he might equally have been referring to hospitals. Others have put it somewhat differently. Brown (1979), for example, has argued that medical leaders striving for professional status sought support from the leaders of industrial capitalism. At the same time, within the dominant ideology, "technological medicine provides the corporate class with a compatible world view, an effective technique, a supportive cultural tool" (p. 228). A powerful coalition of the leaders of medicine and industry thus emerged. Brown stresses a process of mutual legitimation rooted in the dominant ideology of the time.[8] If we look at the introduction of a specific technological device, such as the Roentgen apparatus for producing x-rays, we find common economic interests also emerging.

The historical process by which medicine became dependent on technology, expressed in and reinforced by the increasing armamentarium adopted and deployed and the increasing precision with which sickness had to be described, is not easily summarized. It is clear that it has many elements, just as science has many meanings. Its elements are to be found within medicine itself, in the processes of its professionalization and specialization, which make a differentiated technological armamentarium a vital resource. They are to be found within processes of social and economic reorganization, which required the large-scale and reproducible screening of huge numbers of symptomless individuals. This also led to the need for diagnostic technologies having certain characteristics. A medical equipment industry, based partly on the redeployment of wartime expertise, came into being to provide these technologies. Social change also brought hospital care within reach of growing segments of the population, as health insurance was extended in many countries following World War II. Here emerged the competitive system in which, especially in the United States but now in much of Europe as well, hospitals compete both for patients and for the best physicians. Through a variety of processes, physicians have been led to attach more and more significance to the evidence provided by machines. This is nowhere more evident than in the field of diagnostics. The process by which the x-ray became a central resource of diagnostic practice is exemplary. It shows the interplay of professional and industrial interests in the development of the technology central to diagnostic medicine even today.

The Establishment of Roentgenography

Wilhelm Roentgen's discovery of x-rays, announced at the end of 1895, took the world by storm. In this period a number of physicists had become interested in the stream of particles, later called cathode rays, produced in vacuum tubes fitted with positive and negative electrodes (Glasser 1934). These rays appeared not to penetrate the walls of the tube in which they were produced and so were difficult to investigate. In spring 1894, Roentgen, professor of physics at the University of Würtzberg in Bavaria, became interested in the rays, and in the autumn of 1895, according to the conventional account, he had an idea. Perhaps these cathode rays actually did penetrate the glass tube, but the luminescence by which they would normally be detected had been obscured by light. In November 1895 Roentgen tested his idea and, seemingly by accident, in setting up his apparatus he discovered that the detecting screen was caused to fluoresce while still a meter from the tube. Cathode rays had never been known to travel more than centimeters through air, and Roentgen thought it likely that some other kind of radiation was involved.

Thanks to the success of a rapid series of tests and a cooperative journal editor, on 1 January 1896 Roentgen was able to distribute copies of his article *"Uber eine Neue Art von Strahlen."* The paper dealt with, among other matters, the penetrating powers of the new rays. In his investigation of this effect, Roentgen had looked at their penetration of the human hand. His paper comments briefly on this: "If the hand is held between the discharge apparatus and the screen, one sees darker shadows of the bones against the less dark shadows of the whole hand." An x-ray "photograph" of his wife's hand, showing bones and ring, accompanied the text. Copies were

sent to a number of eminent physicists including, in Britain, Lord
Kelvin and Arthur Schuster, a German-born professor of physics in
Manchester. In Vienna Franz Exner was so impresssed by the article
and, especially, by the photographs included with it that he informed Di*e Presse*, Vienna's leading newspaper, which published
news of Roentgen's discovery on its front page on 5 January 1896.
The story spread through the world's newspapers like wildfire. It
was first reported in England, in the *Daily Chronicle*, on January 6,
under the title "Remarkable scientific discovery" (Posner 1970).
The newspaper article, appearing only five days after Roentgen had
circulated his report, already referred to possible medical uses of
the new technique, commenting that it will be "an excellent expedient for surgeons, particularly in cases of complicated fractures of
limbs, in searching for the bullets of the wounded, etc."

The apparatus required for replication of Roentgen's experiment was reasonably available; anyone with access to a physics
laboratory could try it out. On the basis of the newspaper report, an
English electrical engineer was able to show a roentgen photograph
of his own hand to friends within a day. The medical press rapidly
took the matter up. Still in early January, the Lancet, at first amused,
carried a report from its Berlin correspondent to the effect that
general opinion was that "the new discovery will produce quite a
revolution in the present methods of examining the interior of the
human body." By February 1896, "traumatic, osteomyelitic, and
tuberculous bone lesions had been shown on radiographs at the
Saltpetière and Trousseau Hospitals in Paris. . . . In the United
States, E. B. Frost had photographed a broken ulna on 3 February.
. . . One day after Frost's excellent photograph had appeared in
Science the editor óf the *Journal of the American Medical Association* was
still very sceptical" (Posner 1970).

For some time the apparatus was treated as a toy as much as
anything else. In his catalogue of references and early displays
Reiser includes cartoons, shop windows, satirical magazine verses,
and fairground sideshows. The remarkable powers of the rays
became a matter for popular speculation, and an x-ray photograph
something to be prized by the fashionable man or woman (Reiser
1978, 60–61; see also Pasveer 1989). Physicists, engineers, and
physicians were no less fascinated, and serious investigations were
pursued widely and enthusiastically. Burrows (1986) has found a
pattern of cooperation in the first medical investigations in Britain.
Generally speaking, a "local doctor or surgeon with a suitable case,

usually a needle lodged in the patient's finger or foot, would seek the help of a science professor in the local college or university, and together they would produce a radiograph" (p. 20). It has been claimed that more than a thousand articles were published on Roentgen's discovery in the course of 1896 (Glasser 1934): 163 articles, letters, and notes in *Nature* alone, and many in the trade journals of the electrical industry.

Among those paying careful attention to this explosion of interest was E. W. Rice, a vice president of the newly established General Electrical Company.[9] Rice was soon convinced that a market for x-ray tubes and apparatus existed (Carlson 1984, 494). On 11 March 1896 he asked Elihu Thomson, a scientist-inventor who worked for the firm, to design an x-ray apparatus (a tube plus a high-frequency coil to power it) for commercial sale. Within a month GE was advertising its new product line in *Electrical Review*. Within that same month Thomson had designed an improved double-focus tube and was working on problems of quality, performance, and durability. GE decided to sell Thomson's new tube for $12. By August 1896 the firm had produced a catalogue listing a full range of x-ray products: the new tube, excitation apparatus, interrupters, and other accessories. They were confident that with their existing network of salesmen they would be able to reach the market they foresaw among doctors, hospitals, and scientific laboratories.

General Electric did not have this emerging market to itself. Since Roentgen was not interested in patenting his discovery, refusing various financial inducements to do so, the apparatus could be freely produced. In Britain three firms, established producers of scientific instruments and microscopes, started manufacturing roentgen tubes in the course of 1896 (Tunnicliffe 1973). Within the year a comprehensive range of x-ray apparatus was available in Britain (figure 1.2). On the basis of advertisements placed in the *Electrical Review*, Carlson estimates that by December of that year at least eight firms were offering x-ray apparatus for sale in the United States. With remarkable speed, a thriving and dynamic x-ray industry came into being. Physicians were among its most enthusiastic customers.

In May 1896 Sidney Rowland, a 24-year-old medical student who in February had been commissioned by the British Medical Association to investigate the clinical uses of x-rays, published the first issue of the world's first "radiology" journal. *The Archives of Clinical Skiagraphy*, after the Greek word for shadow, became the *Archives of*

Figure 1.2
Apparatus for Roentgen diagnosis, 1897. From Walsh 1897, reprinted by
permission of the Library, Wellcome Institute for the History of Medicine.

the Roentgen Ray the following year. In December Walter Cannon,
later one of the world's leading physiologists but then a medical
student, asked his professor (Bowditch) for a research problem
(Harvey 1981, 43). Bowditch suggested that he and a fellow student
use the newly discovered rays to test a theory about the nature of
swallowing. Using a fluorescent screen to obtain a "real-time"
image, Cannon and Moser radiographed a dog while it swallowed
a pearl button. Their paper was published in the first issue of the
American Journal of Physiology in 1898. The x-ray was becoming a tool
not only of the field surgeon locating bullets but also of the
physiological researcher.

It is important to note that in the early years the practice of
roentgenography, or skiagraphy, was not limited to a specialized
class of medical practitioners. Those offering a diagnostic service
around the turn of the century were a motley crew. Alan Campbell
Swinton, who made the first x-ray picture of the human body in
Britain—noted with interest by *The Times*—was a self-taught electrical engineer. Like others with engineering and other technical
backgrounds, he set up in private practice, offering a "lay" roent-

genographic service, a concept that would later be challenged by the medical profession (Burrows 1986, 24–27; Larkin 1983, ch. 3). In the first years there was no control over who might take, or interpret, roentgenographic pictures. Well into the twentieth century many hospitals employed lay "radiographers," without medical supervision, to make as well as to interpret x-ray films. For that to happen, of course, the diagnostic value of the roentgenograph had first to be appreciated, and this was itself a complex and protracted process (Pasveer 1989).

When the Roentgen Society was formed in London in 1897, it deliberately opened its doors to anyone with a scientific interest in x-rays. Apparatus makers (the budding x-ray industry), photographers, and physicists were among its most active members, and the first president was Silvanus Thompson, a physicist. A survey of the members of the American Roentgen Ray Society in 1910 showed that there, too, many had started working with x-rays in 1896 or 1897 "as physicists, engineers, electricians, photographers or in other technical capacities," but "had thereafter gone back to school and earned their MD degrees specifically for the purpose of qualifying as a radiologist" (Brecher and Brecher 1969, 109).

Despite the explosion of interest in the new rays, change in routine diagnostic practice was more gradual. Here is how A. E. Barclay (who took up the study of medicine in 1899 and a post as radiologist in Manchester in 1906) described his time as a student at the London Hospital, where x-rays were introduced in the course of 1896: "Few of the surgeons yet fully trusted x-ray plates and I remember a very heated argument between two of the honoraries, one of whom was an enthusiast for their use, the other maintaining that his clinical examination was more reliable. . . . Sometimes, hopefully rather than expectantly, x-ray plates were asked for of the chest, kidneys, bladder, or even the spine. . . . For practical purposes, however, the x-ray diagnostic service was confined to fractures and metallic foreign bodies, particularly the many cases of needles in the hand that came from the tailoring trade of Whitechapel" (cited in Burrows, 79–84).

It seems clear from Burrows's detailed account that while many hospitals in the British Isles obtained x-ray equipment within the first year or two, its use varied widely. This probably had little to do with concern over the safety of the technique. Although reports that x-rays could destroy life and damage the skin were appearing with increasing frequency by 1900, and many investigators and "radiologists" suffered burns due to inadequate protection, physicians were inclined generally to be dismissive of their harmful effects.

At the London Hospital and at St. Bartholomew's, two leading London teaching hospitals, x-ray equipment was placed in established electrotherapeutic departments, directed by physicians. At another leading London hospital, St. George's, x-ray pictures were made by a nonphysician until 1907. Falling into the former category was the Royal Infirmary in Glasgow, where the aging Lord Kelvin had been an enthusiastic recipient of one of Roentgen's preprints. Glasgow Infirmary established a well-equipped "Roentgen laboratory" in June 1896. The situation at the Pennsylvania Hospital, a leading U.S. institution, was comparable with that at St. George's. Although an x-ray instrument was purchased in 1897, it was not used. Even in 1902 it was rarely used, and then more out of curiosity than for patient care. It was a decade before things began to change (Howell 1986). The Pennsylvania Hospital only then appointed a full-time roentgenologist and established a separate x-ray unit. The number of patients examined then began to rise steadily.

Howell points out that the experience of World War I, in which x-rays proved invaluable in the field, provided persuasive evidence of the technique's value. This was true in many centers. In Glasgow, the number of patients x-rayed rose from 1400 in 1901 to 2730 in 1906 to 5725 in 1914. A new x-ray department was opened in 1914— just in time. The war brought a huge increase in work, reports Burrows: 7000 radiographs were made in the first seven months of 1916, with sixty to eighty injured servicemen needing to be handled in an afternoon (p. 119). It was this expansion, plus the large numbers of people, both medically trained and untrained, who acquired field experience of x-ray equipment, that forced consideration of the organization of the service on the leaders of the incipient profession.

The first medically qualified roentgenologists were all too aware both of the way in which treating physicians and surgeons viewed their place in the medical scheme of things and of the problem posed by medically unqualified personnel. The fight to establish their own specialized medical role would have to be carried out on two fronts, consonant with the general picture that Starr has painted of the process of specialization. In the larger British hospitals a trained doctor was generally in charge of x-ray work. However, as A. E. Barclay recollected, that person was generally regarded as a technician. "It was only slowly that his colleagues came to recognise that the radiologist's constant experience in interpretation made his opinion on x-ray plates of real value, yet by 1914 there were still very few who had attained anything approximating to consultant relations with a their colleagues" (cited in Burrows 1986, 179). In a fighting presidential address to the Röntgen Society in 1916 Thurstan Holland argued that the time of the technician, skilled only in the making of plates, was over: "The real essential for an x-ray expert [is] interpretation. The good plate can now be obtained by anybody with modern apparatus with very little training [but] interpretation can only be done by a medical man of unusual professional attainments" (quoted by Burrows 1986, 180).

It was in 1917 that physician-radiologists in Britain took steps toward professionalization within medicine. In that year a new organization was formed, initially named the British Association for the Advancement of Radiology and Physiology, "to promote the advancement of Radiology and Physiotherapy on scientific lines under the direct control of the medical profession" (Burrows 1986, 182). This body entered into negotiations with the universities with the objective of establishing a university diploma in radiology.

By 1913, according to Brecher and Brecher (p. 211,) while the United States boasted a "few hundred" radiologists, many of them were also in general practice or in some other medical speciality. This was still true at the beginning of the 1930s, though the numbers of radiologists had risen. The AMA Directory for 1931 listed 1005. Leading radiologists in the United States complained that the standard was low, a failing that they felt encouraged other specialists to believe they could easily pick up the smattering of necessary knowledge. Through the 1930s, nevertheless, the status and professional independence of radiology within medicine were becoming established. By 1930 London University had established a chair in radiology, and a number of British universities offered postgraduate diplomas. In the United States, by the end of the 1930s, 75 percent of physicians who provided radiological services limited their practice to this field. New applications were being developed, including radiographic examination for cancer of the breast. Gershon Cohen's original paper on the early diagnosis of breast cancer appeared in 1937 (although development of a sound and reproducible "mammographic" technique took place only in the early 1950s) (Evans and Gravelle 1973). Mass chest radiography was developed in the 1940s, and after the war mobile units became familiar arrivals as children were regularly screened for tuberculosis (figure 1.3).

Rapid growth in obstetric radiology also took place between the mid 1940s and the mid 1950s. By 1955, in the United Kingdom, about one woman in seven subsequently giving birth to a live baby was radiographed during her pregnancy. The scale of radiology had grown enormously, and it had become a more or less universal feature of hospital practice. All general hospitals in the United States with more than fifty beds, and more than three-quarters of those with fewer than fifty beds, had x-ray equipment by 1951. A U.S. household survey of 1960/61 concluded that more than a quarter of the total population received a chest x-ray during the year.

By this time, however, certain worries regarding the use of x-ray's were surfacing (Blume 1981). In an important series of papers Stewart and her colleagues at the University of Oxford were beginning to amass epidemiological evidence suggesting that exposure to x-rays posed a risk of subsequent cancer to unborn children. In Britain the proportion of all live births previously x-rayed in utero began to fall, and 1957–1960 has been termed the period of "greatest radiological caution." Technological improvement, which had been constant in this area—faster films, higher-voltage techniques with intensifying screens—was leading to major reductions in dosage. It was reasonable to suppose that the risk was being reduced. However, a 1968 paper by Stewart and Kneale comparing cancers in a series of birth cohorts at age ten found that there was no clear effect on the incidence of cancer attributable to obstetric radiography. In other words, looking at the total population, there was no clear trend toward safer obstetric radiology.

By the 1970s almost all modern hospitals were carrying out roentgen diagnosis. Despite the evidence that might justify caution, use of x-ray equipment even in the obstetric area was not declining further. A British study of hospital practice in 1974 showed that, on average, 22.7 percent of pregnant women were being x-rayed. In one hospital studied, the proportion rose from 30 percent in 1953 to 39.7 percent in 1955, then fell to 11 percent in 1961 and rose to 23 percent by 1974 (Carmichael and Berry 1976). These authors commented that "where a patient attends a hospital with readily available x-ray facilities, these facilities will be used." In other areas

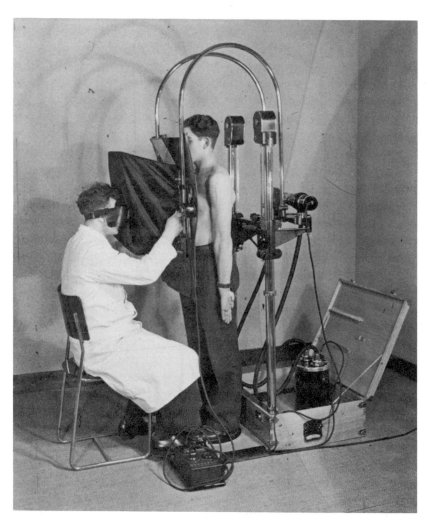

Figure 1.3
Portable equipment for x-ray chest examination, c. 1932. Reprinted by permission of Philips Medical Systems.

of diagnosis, too, studies suggesting if not the harmfulness then at least the limited value of various radiological procedures have been without effect (Potchen 1981). Overuse or inadequately selective use of radiological procedures has frequently been attributed to current beliefs regarding the diagnostic process. Impressed by the "hardness" of machine evidence, anxious to seem to have left no stone unturned, physicians typically order all relevant tests, giving little thought to the marginal value of "yet another" investigation. Given the inherently technological diagnostic culture that results, small wonder that diagnostic tests of various sorts (pathology tests, nuclear medicine procedures, laboratory tests, radiological investigations) have been among the fastest-growing aspects of medicine (Redisch 1982, 140–141).

Of course, some diagnosticians are aware of what is going on and are duly critical. Radiologists have been criticized by the professor of radiology at Cambridge University for their "unscientific" practice in, typically, looking to collect all possible data on a patient. The properly scientific way of proceeding, he argues, would be "forcing a critical diagnostic pathway" through the "investigative jungle" by formulating and testing hypotheses (Sherwood 1979). This alternative would not only be more scientific, it would at the same time represent a far more rational use of the various available techniques of investigation. Here is what a professor of medicine at Yale has to say regarding the seductiveness of machine evidence:

The worst problems come when the doctor fits the patients to his skills, something which is true of all professions. A woman comes to a gastroenterologist and gets a sigmoidoscopy, a barium enema and a high fiber diet. Going to a gynecologist, she runs the risk of laporoscopy and of losing her uterus if she contines to complain. We must get back to training students and residents to look at the patient rather than to look simply at the data base. In general, I refuse to hear about the data base or to look at it or pay any attention to what the x-rays or other studies show until I have talked to the patient. I know that the minute I see the x-rays of the patient, before looking at the patient or before working on him, I will fit the patient's story into whatever the x-rays or other images are showing me. (Spiro 1981, 45)

Yet the demand for diagnostic data has remained insatiable. Maintained by powerful incentives, it has at the same time helped sustain the process of technological innovation that has led to the machines described at the start of this chapter. Nowadays imaging apparatus is generally located in a centralized radiology department, which (in the United States at least) tends to be "responsible for all imaging procedures" (Hamilton 1982, 4). Other departments of the modern hospital may also deploy it on a more limited scale. As one American study has put it, "diagnostic imaging equipment can also be found in the emergency room, in the cardiology department, and in the department of obstetrics and gynecology" (ibid.). For the radiological profession, however, this instrumentation is crucial: the source of status and function within medicine. Deriving from a gradual specialization in the use of Roentgen's discovery, radiology has become a major instigator of and customer for new imaging technology and for the new insights that it promises.

Understanding Technological Change in Medicine: The Case of Diagnostic Imaging

The early decades of roentgenography were marked by processes of organizational change in two areas. First, a role had to be created within the medical system. We have seen that the medical profession made space for the new specialty only gradually. Here the story is one of prolonged struggle—lasting well into the 1930s—for radiology's professional status. Second, manufacturing industry also adapted itself to the new discovery. This was accomplished rapidly, as we saw. Within a year of its discovery, manufacturers made the means of generating an x-ray image widely available for a relatively modest capital outlay. Manufacturers and users alike seem to have been enthusiastic about the possibilities of improvement to the basic technology. Consider Barclay's recollection of the continuous improvements to x-ray apparatus after 1918: "Coils were replaced by close-circuit transformers, with mechanical rectification and much higher output. The loose tangle of wires gave way to an orderly arrangement of high tension conduction tubes provided with change-over switches. Gas tubes, unsuitable for current from a transformer, were superseded by hot cathode tubes. Films replaced glass plates. . . . The greater efficiency of apparatus, increased sensitivity of films and improved intensifying screens made it possible for exposures to be measured in seconds and even fractions of seconds" (Barclay 1949). Yet technological innovation in roentgenography today is a different process from the processes that Barclay may have observed. The difference is intimately connected with the coming into being of the organizational field centered on x-ray technology.

Elihu Thomson, whose work had been the means of GE's entry into the new market, in the course of 1896 became interested in the medical applications of the technology. Perhaps it helped that, having broken his leg in a bicycle accident, he was himself able to profit from roentgenography. Thomson was often asked to speak to medical organizations, and he must have learned a good deal about how x-ray devices were being used. Yet it appears that these experiences had little influence on his further work on x-ray technology. Carlson's study of his work shows how Thomson continued to focus on the issues that had previously concerned him, such as more effectively regulating the vacuum. Carlson suggests this continuity is in need of explanation and, further, that the explanation might be found in the fact that Thomson saw the "electrical engineering and physics communities as his main audience; they, not the medical profession . . . could provide him with the approbation that he desired" (Carlson 1984, 525). The fact that Thomson soon moved on to other research attests to the plausibility of this contention.

Yet there is more at stake here than one scientist's intellectual interests. The lack of significant effect from Thomson's interaction with medical users of x-rays is also reflective of commercial and innovation strategies in the early years of the industry. Manufacturers addressed themselves to a relatively ill-defined group of customers. Perhaps united principally by their fascination with technique, those purchasing and using the x-ray tubes that Thomson designed for GE must have been a fairly heterogeneous group. Manufacturers had not yet learned to value their experience as a source of input to development work.

For industry as well as medicine World War I was a watershed. On the one hand it made the potential of the new diagnostic technology apparent to physicians, while on the other industry began to see that large markets could be in the offing. It was the sheer volume of sales to the military during the war that persuaded General Electric (which had ceased to produce x-ray equipment around 1905) that they should become a "full line x-ray equipment supplier" (Reich 1985, 91).

It was in the context of that decision that William Coolidge led work at the GE research laboratory to develop improved x-ray tubes. He soon realized that the tubes were so erratic partly because increasing gas pressure affected their operation. By substituting tungsten for platinum as the positive electrode, Coolidge developed a tube that could be operated at high temperatures, reducing this problem. The laboratory started manufacturing these on a small scale, selling 6500 in 1914 (Reich 1985, 89). Subsequently, after considerable further work, Coolidge developed a still more effective tube. Medical radiologists were extremely enthusiastic about this "Coolidge tube."[10] Many further developments followed (see figures 1.4 and 1.5). Soon afterward came the Potter-Bucky grid, an invaluable device for eliminating image blurring due to the production of secondary electrons within the body (Brecher and Brecher 1969, 205–210).

However much they may have welcomed technical advance, there is evidence that medical users did not view these processes of innovation with unalloyed enthusiasm. In an interesting contribution to discussion of "the professionalization of radiology in America," A. U. Desjardins of the Mayo Clinic linked the difficulty in establishing secure professional standards with the ready availability and rapid development of the roentgen apparatus (Desjardins 1929). "The trend of development in the technic of roentgenography,"

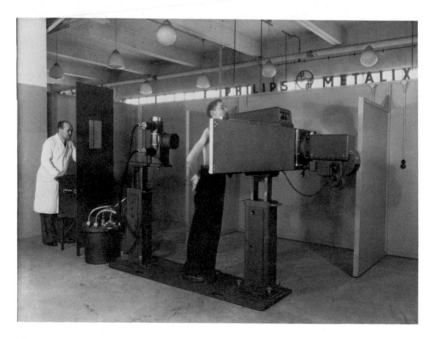

Figure 1.4
X-ray equipment around 1950, showing fluorescent screen, camera and image intensifier. Reprinted by permission of Philips Medical Systems.

Figure 1.5
The internationalization of markets: portable x-ray equipment similar to that shown in figure 1.3 in use in China in the 1930s. Reprinted by permission of Philips Medical Systems.

Desjardins wrote, "has been marked chiefly by the constant endeavor of manufacturers to produce generators the operation of which would be so simple that an increasing number of physicians would be impelled to purchase them, regardless of their lack of that extensive body of special medical and physical knowledge necessary to take full and intelligent advantage of the possibilities of roentgenography." Associated with this was the commercial practice of the new industry: "This trend toward simplified construction, ease of manipulation, and standardization of technical procedures, accompanied by a tremendous extension of the credit system in commercial transactions, a corresponding expansion of advertising, and the development of high-pressure salesmanship as a means of increasing the volume of sales, has had much to do with the wide dissemination of roentgen-ray apparatus not only in legitimate medical circles but also among the quacks and outlaws of the borderlands."

In the preprofessional period, industrial suppliers were directing their innovative activity to a broadly medical and unspecialized audience. By so doing, as Desjardins suggests, they were wittingly or unwittingly hindering the radiologists' efforts at securing professional control. Gradually, with professionalization, users became restricted to medically qualified radiologists. Henceforth it would be this more clearly demarcated group of customers that would have to be satisfied. The innovation process changed, as industry and their radiologist customers came to play symbiotic roles in the process of technical improvement that has marked roentgen technology throughout its history.

Although the emphasis vis-à-vis the medical profession may have been on expertise in interpretation, this was not, in fact, at the expense of technical knowledge and ability. Professionalized radiology inherited the fascination with technique, with equipment, that its proto-practitioners displayed. Partly, as Brecher and Brecher put it, the early radiologists sought challenges in the manner of technologists. Once localization of foreign bodies and of fractures became commonplace, "the more fascinating challenge which the early radiologists faced was how to extend the boundaries of their art into wholly new regions—the brain, for example, and the gastrointestinal tract. To meet this challenge they had to devise new techniques" (Brecher and Brecher 1969, 110).

Innovation was coming to depend on a common and symbiotic commitment on the parts of both manufacturers and the profession of radiology. This was perfectly expressed by an eminent American radiologist, writing in 1945: "Some critics have contended that radiologists should now leave all of their technical problems to the engineering staffs of the commercial manufacturers. It seems safe to assume that eventually our radiological tools will reach such a degree of perfection and standardization as to allow us to ignore the details of their construction and operation, devoting ourselves exclusively to the medical phases of our work, but that day is not yet here" (Hodges 1945; see also Hiddinga 1991). Nor is it yet, forty-five years later. With growing interdependence, the perspectives on innovation of an increasingly oligopolistic industry and an increasingly professionalized radiology were becoming more and more attuned to one another.

Until the 1950s, x-ray technology was the only imaging technique available for diagnostic purposes.[11] Its success led to the development of numerous specialized applications in neuroradiology, in coronary care (angiography), in relation to the pulmonary system (bronchography), and in mass screening programs such as for breast cancer. Many of these techniques have become associated with highly specialized roentgenographic instrumentation. The history of roentgenography is one of more or less continuous but until recently incremental technical change.[12] In this book we shall be concerned with the development of imaging techniques radically different in their underlying technology from roentgenography.

The radiological profession clearly had—and has—a firm commitment to technological innovation within its area of practice. This is due partly to the fact that, through their practice, radiologists become aware of desirable improvements to their equipment. Yet at the same time professional interests may serve to set limits to the innovation process. After all, while innovation is limited to incremental improvement to existing devices, radiologists' own expertise (essentially in the *interpretation* of x-ray plates) remains valid (Barley 1986). For many decades this is how it was. But from the end of the 1950s the innovation process began to change. New possibilities of diagnostic imaging emerged, based upon the deployment into medicine of wholly different technologies. These new technologies provided a radically different semiology, thereby throwing the existing interpretative skills of radiologists into the balance. Control of new imagery can be all-important, and in this book we shall be concerned with the interaction between "control" (in this sense) and the development of new techniques for diagnostic imaging.

A modern imaging technology, such as the PET scanner in Sochurek's *National Geographic* story, can be described as having a "career," and this notion will be developed as part of the theoretical scheme of chapter 2. The emergence, and the subsequent careers, of such technologies is intimately bound up with central interests both of radiologists and of the x-ray industry. In trying to understand the emergence of diagnostic imaging technologies—and in thereby posing some of the larger questions with which this book is concerned—the complementarity, or perhaps symbiosis, of these interests will be of central significance. In chapter 2 we shall look theoretically at why this is. In subsequent chapters we shall examine the careers of four new types of diagnostic imaging devices, four new ways of viewing and judging the secret places of the human body. We shall see how these artefacts were conceived, designed, and shaped—how the image that they (and those who conceived them) offered was taken up, approved by medical practice, and began to reshape that practice. As we do so, we shall see how the process of innovation was structured by what I shall term an "interorganizational structure"—by the central interests binding together institutions providing radiological insight and the industry producing the devices for generating it.

2

With Subtle Skill and Elaborate Research

Introduction

How should we try to understand the development of new diagnostic imaging devices? In chapter 1 we saw that the establishment of modern radiology entailed adaptation on the parts of both hospitals and manufacturers. With the professionalization of radiology there emerged a symbiotic interest in technological advance on the part of both the professional users and the industrial producers. Manufacturers gradually shifted their efforts to improve roentgenographic devices, focusing on the more specific demands of an articulate research-oriented user community. Sustained by a growing physician demand for "hard" diagnostic information the "organizational field" centered on hospitals became committed to the deployment of increasingly sophisticated diagnostic devices. The shared commitment to innovation of radiologists and the industry supplying them found support in the values and the funding structure of the hospital system. After the Second World War the shared commitment to new technology became still more firmly embedded, as recently demobilized technical experts explored the relevance of their knowledge for the development of improved medical devices. This was nowhere more true than in the field of diagnostic imaging. But what does the innovation process entail? How does the vision of a new imaging principle take shape and how does it become integrated into medical practice? The objective here is to develop an appropriate analytical scheme, based on the notion that we can only understand innovation in diagnostic imaging relative to the structure of relations between radiologists and the industry producing roentgenographic equipment.

This chapter surveys some of the established perspectives now used in trying to explain technological innovation. It begins by reviewing some of the economic literature dealing with technological innovation, merely hinting at its specific application in the area of medical technology. The sociological position, which is in a sense a principal point of departure for the perspective of this book, is presented in somewhat more detail, using another historical case study (the electrocardiograph, invented just a few years after the x-ray machine). The argument is offered that a number of convergencies between economic and sociological approaches are beginning to take place; the perspective of this book is located there. The chapter concludes by presenting the analytical scheme to be used in developing the case studies of chapters 3 through 6.

Economists' Problematizations of the Innovation Process

Economists have debated questions relating to the sources and nature of industrial innovation for at least fifty years, though the roots of the discussion go back to the very beginnings of economic thought (see Freeman 1977 for a review). In some of his work on the subject, Schumpeter argued that the driving force of innovation was to be found in the personality of the entrepreneur, while elsewhere he emphasized the importance of monopoly as a stimulus to innovation. More recently economists have looked in some detail into how innovation takes place, into the relation between technological achievement and commercial success, and into the effects of government policies on the innovation process. Economists have sought systematically to understand the speed, the nature, and (though only in a very limited sense) the direction of technological innovation.

For long after empirical work began in this area, in the 1950s, the process of innovation itself remained hidden—a black box. Studies showed a high correlation between research and development expenditures on the one hand and national economic growth on the other, and this seemed to underpin a certain faith in the inevitable profitability of technological innovation. The economic studies of this period abjured any concern with the content of technology. The notion of the direction of its development was conceived, if at all, in terms only of the relative inputs of labor and capital needed in the production process. While a technology might thereby be conceived of as labor saving, there was little or no connection with work in the history of technology. Given reasonably auspicious external circumstances, research would lead to innovation, and innovation would be profitable. Gradually this assumption came to be questioned, giving way ultimately to a dispute centering on the relative importance of demand, or market forces on the one hand, and the supply of knowledge on the other: the market pull/technology push debate. Jacob Schmookler was a major proponent of the view that the source of technological innovation is to be found in the market (Schmookler 1966). According to him, economic forces (the expectation of profit) determine both the extent to which innovative activity (which nowadays means organized research and development) takes place, as well as the the likelihood of this activity being addressed to one class of goods rather than another. Thus market demand will stimulate entrepreneurs to try to develop one sort of product or another (and influences how hard they try); knowledge, science, is of importance only in suggesting the best means to set about this development work.

We can imagine what a Schmooklerian explanation of turn-of-the-century medical technology would look like. Diagnostic x-ray (and the ECG) emerged at a time of rapid expansion of the hospital system, a time when the middle class began to make use of hospitals. The ripples of this rapidly growing market for hospital equipment were felt even within the basic research system. It was no coincidence, the Schmooklerian would claim, that just at that time so many new journals of medical science were founded. (As previously noted, Walter Cannon was able to publish his x-ray studies on swallowing in the first issue of the *American Journal of Physiology*.)

Among those who have opposed this extreme emphasis on the determinacy of market forces has been Nathan Rosenberg. In a number of important works Rosenberg has produced many historical examples of human needs that, even when clearly expressed through the market, failed to give rise to the inventions that would satisfy them. This Rosenberg attributes to the constraining influence of available knowledge "growing at uneven rates among its component subdisciplines." For Rosenberg economic forces can only operate within such a framework of scientific and technical constraint. Rosenberg's work, like that of other historically inclined economists, has also drawn attention to *what* exactly gets invented—to the attempts of producers to improve those aspects of their products that limit their operating efficiency (Rosenberg 1976). If we take Rosenberg's position seriously, we would have to try to make sense of what exactly people were trying to invent, the ways in which they were trying to improve their products. It would not be enough to argue that so many resources were devoted to improving a black box that in this case is the x-ray tube.

Few are now inclined to argue as to the relative importance of market forces and the supply of scientific and technical knowledge. They are recognized as complementary influences on the course of innovation, with rather different significances (Mowery and Rosenberg 1979, Kamien and Schwartz 1982). In a much-cited review of the literature Mowery and Rosenberg write, "any careful study of the history of an innovation is likely to reveal a characteristically iterative process, in which *both* demand and supply side forces are responded to. Thus, successful innovations typically undergo extensive modifications in the development process in response to the perception of the requirements of the eventual users on the one hand, and, on the other, in response to the requirements of the producer who is interested in producing the product at the lowest possible cost."

Much more important today is how innovation takes place: the social and reasoning processes through which decisions are made and ideas take shape. This leads inexorably to a new concern with the direction of technological change. Why, for example, did x-ray apparatus evolve in the ways that it did and not in other ways? How was it decided that hot cathode tubes should replace gas tubes and films replace plates? It is the question, not the answer, that is relevant here.

Prior studies offer reason to believe that there are circumstances under which we may expect the user, the customer, to play an important role (von Hippel 1976, von Hippel and Finkelstein 1979). When Elihu Thomson busied himself with improvements to Roentgen's original apparatus, he was little interested in building on medical experiences with its use, despite his personal interest in medical applications. When Coolidge invented the hot cathode tube twenty years later, the enthusiasm of radiologists and a general sense of GE's image (or responsibilities) seem to have been the dominant considerations in marketing it. The suggestion of chapter 1 was that the process of innovation in x-ray equipment became one such as von Hippel characterizes, but gradually, with the emergence of the interorganizational structure in its modern form.

On the basis of daily practice the radiologist sees how improvements can be made in the equipment with which he or she works. But only with emergence of the modern interorganizational structure did these perceptions become a vital input into industrial development work.

I think it should already be possible to see something of what is happening to the economic study of innovation. "Perception of the requirement of the eventual users," users "trying to persuade manufacturers": economic thinking in this area is becoming much more "behavioralistic." In the evolutionary perspective of Nelson and Winter, which has attracted a lot of attention from more sociologically inclined students of innovation, this tendency is much more pronounced.

Nelson and Winter have argued that we should look at the process of technological innovation, and that firms look at innovation possibilities, more in terms of "backing and forthing" between demand and supply side considerations (Nelson and Winter 1977). At any given time the firm is likely to be faced with a choice of ways forward, all—and this is crucial—beset by uncertainties.[1] Choice of research projects can only be probabilistic, based upon a set of assumptions as to consumers' relative demand for different sorts of products or improvements, on the state of knowledge (how difficult different sorts of innovations will be), and on criteria governing how much in total can be made available for research and development. Most important, Nelson and Winter try to develop a model in which trial, the establishment of preferences by customers, customers' experiences and selections, interact with and partially shape what producers choose to do. This is how Nelson and Winter stress the interaction between the decisions about innovation that supplier firms take on the one hand, and what amounts to the diffusion of innovations among customers on the other:

Despite a tendency of some authors to try to slice neatly between invention, and adoption, with all of the uncertainty piled on the former, one cannot make sense of the micro studies of innovation unless one recognizes explicitly that many uncertainties cannot be resolved until an innovation actually has been tried in practice . . . analysis of the ways that innovations are screened, some tried and rejected, others accepted and spread, must be explicitly dynamic. We propose the concept of a "selection environment" as a useful theoretical organizer. Given a flow of new innovations, the selection environment . . . determines how relative use of different technologies changes over time. The selection environment . . . feeds back the influence strongly of the kinds of R&D that firms and industry will find profitable to undertake.

Much of the emphasis these days, then, is on the firm as an actor, developing strategies to meet the uncertain world with which it is faced. In Nelson and Winter's work we have two sorts of innovating organizations in interaction with each other. For the firm, the signals it receives from the "selection environment" (the organizations using its products) are of importance in determining its innovation strategy. (Another consideration is said to derive from the extant "technological paradigm." According to Dosi's extension of Nelson and Winter's work, there are also technical rules embodying strong prescriptions as to which directions of technological change to pursue (Dosi 1982).) Firms in models of this kind are practicing what has become known as "bounded rationality"; trying to behave rationally, they are constrained by lack of knowledge, including of the consequences of their own actions.

According to this perspective, then, empirically to understand the innovation process in a given area—such as medical devices— we have to look simultaneously at producers and at their market. The medical devices industry is very diverse with a few giant multinationals controlling very large shares of the market (GE, GEC/Picker, Philips, Siemens, Toshiba among others), but also with a large number of small companies, some of them specializing in one product.[2] A recent study of the industry in Britain found it to be characterized by a high degree of takeover and merger activity and a high rate of entry of new firms (Hartley and Hutton n.d.). The history of the production of x-ray equipment in Britain is typical with the first three firms gradually becoming tens of firms by the "exuberant" 1920s and 1930s and then slowly devolving into just one firm—GEC—by the 1960s (Tunnicliffe 1973).[3] From the point of view of these firms the health care system is a market. Moreover it is a special kind of market because it is also an industry. It is an industry providing a certain range of services, and the goods that it buys (the beds, bandages, and drugs just as much as the sophisticated new devices) are the tools of its trade (thus comparable with the market that banks form for computers or armies for guns). It is a very large industry, and taken together there is little competition for the services it provides. We saw in chapter 1 how access to these services gradually came to be seen as a right to which all in affluent western countries may lay claim. It is an industry that is quite differently organized in different countries: privately run (as for the most part in the United States), or provided by or under the auspices of the state (as in much of Europe). This internal organization of the health care system is important for innovation processes. There are economic reasons to suppose that a rather competitively organized sectorial organization (like the American medical system) will provide bigger incentives to innovate in "processes of production" than one that is monopolistically organized (Arrow 1962, Kamien and Schwartz 1982). In other words, one ought to expect the American health care system to be more innovative, more receptive to new technology than European systems in which the state plays a dominant (quasi-monopolistic) role.[4] Second, there are also economic reasons to suppose that because there is no substitute for the service provided by the health care system, there is little or no systemic incentive to seek to bring costs down (Kamien and Schwartz p. 43).[5] Since governments in Europe and the United States are now trying to do precisely that, the stage is set for conflict. In any event, economic theorizing would suggest innovation in medical technology to be aimed at something other than cost reduction, as is generally agreed to be the case.

Hartley and Hutton's study, apart from giving a picture of the British medical devices industry, also characterized the "typical innovation." The study confirms empirically what we had supposed on theoretical grounds about the nature of innovation. Since there is little competition from outside the formal health care system, there should be little incentive to try to cut the costs of provision of services: it was unlikely that innovation would be aimed at cost cutting. Hartley and Hutton confirm that the supplier industry in a sense contributes here, since "typically, a firm preferred customers to assess its products on the basis of quality and technical capability rather than price." Moreover the greater part of the development work firms were doing could be seen as "the steady improvement of existing products" rather than wholly new products. This finding is suggestive of the sorts of questions we shall need to raise later. What counts as "improvement"? How do the firms know in what way—or better in what direction—to change their products? With what sorts of interests, and how, do firms in an industry like this set about their development work, whether to create new products or to improve their existing ones?

Interesting convergences are now beginning to occur between the study of economic behavior, including innovation, on the one hand and organizational studies on the other. Thus, recognizing that a firm can often choose to relocate its own boundaries "upstream" or "downstream" in its production and marketing, Williamson thereby problematizes the boundaries of the firm. For example, many of the components used in production could be bought from a supplier—or alternatively the firm could make them itself (perhaps by taking over the supplier). Williamson argues that a firm can best cope with some of the uncertainties it faces by "integrating" functions previously performed for it outside (Williamson 1975, 1981).[6] For example, interdependent units can respond to unforeseen contingencies in a more coordinated way, negotiations can be better controlled. Williamson's ideas have recently been applied to the R&D function and to innovation (Phillips 1979, Granstrand 1979) and have been found useful. The factors characterizing the conditions under which internalization will be more efficient apply quite significantly to the R&D process, according to Phillips. Similarly Granstrand finds that much of his data relating to industrial innovation in Sweden is consonant with the notion that "internal organization may be superior to market organization with respect to R&D and innovation" (Granstrand 1979, 358–359). Successful innovation, he found, was often associated with the formation of structures crossing organizational boundaries. Granstrand concludes that "quasi-integrated forms in between pure forms of internal organization and market organization" will be most suited to innovation in many areas of technology (360–361).[7]

Naturally there are many questions that even the most "behavioral" economists do not ask about technological innovation, either because they are not relevant within the theoretical framework being used or, conceivably, because they lie beyond the scope of economic theorizing. A new product has a relatively unproblematic significance for economics, being essentially equivalent to the set of economic parameters that characterize it (e.g. its cost of production, the amount it can be sold for, and so on). But consider the notion advanced by Lawrence, which was discussed in chapter 1: the notion that nineteenth-century medical technology functioned in part as a kind of Trojan horse permitting the advocates of a scientific medicine to overthrow the bastions of those who opposed them.

The idea that technology might have an important symbolic function, or explanation of the astonishing interest that Roentgen's discovery aroused (itself relevant to the *constitution* of the market that developed) are ideas that do not belong within the scope of economics.[8]

Of Hedgehogs and Foxes: Sociologists' Problematizations of Technology

There is nothing new about the notion of a sociology of technology. Indeed, it is one that seems constantly to be being reinvented in terms of one or another sociological perspective. There is a thin thread of sociological inspiration in the history of technology, linking for example Lynn White's classic study *Mediaeval Technology and Social Change* of 1962 to Thomas Hughes's *Networks of Power* of 1985. There are relevant mainstream sociological interests, including studies of diffusion of innovations (which go back to the 1950s). Most important in this context are sociological studies of organization, for here technology has often been seen as a crucial determinant of organizational structure, as the independent variable upon which organizational behavior depends. What organizational theorists typically have meant by technology is in fact limited to the technology used in production: they have not generally interested themselves in accounting for the development of technological artefacts.

A more recent approach that has come to be known as the "new sociology of technology" (see Pinch and Bijker 1984, MacKenzie and Wajcman 1985, Bijker, Hughes, and Pinch 1986) shares a good deal of the theoretical starting point of this book—most notably the notion that technological artefacts embody the circumstances of their constitution. Made by men driven by certain aspirations, ambitions, or interests, there is nothing inevitable in the form that a given technological vision takes. It is a contingent process. One of its essential assumptions is a problematization of the "meaning" of a technology. The notion that a technological artefact has multiple meanings is the starting point for the analysis. Let me make this clear with a simple example.

In her book on innovation in health services, Barbara Stocking takes as one of her examples the neonatal intensive care unit (Stocking 1985). What is a neonatal intensive care unit, and how did it originate? What it is is simple enough. It is a means of caring for the low birth weight, often premature, newborn for its risky (neonatal) period so that it can be returned to its mother with a much better chance of survival. The story of its development seems equally straightforward. If we therefore asked a well-informed pediatrician what a neonatal intensive care unit was, and how it had been conceived, this is the kind of answer we might receive. But what happens when we ask someone with a quite different relationship to this set of procedures, techniques, and organization—what happens when we no longer "privilege" the pediatrician's account? What is the meaning of the neonatal intensive care unit for the mother whose baby lies there? Is it the last hope of saving her baby, a wonder of modern science? Or is it an inhuman prison, depriving her of the closeness to her baby which she craves? What is it for the hospital administrator: a balance sheet, a staffing problem, a service to the community? When we admit that it is all of these things, and that if we want to understand this artefact we have to recognize that, then discussion becomes a little more complicated. If we stress that a technology is all the things that those involved with it say that it is, that its significance is the *set* of all its significances, that there is no reason to give more weight to one point of view than another, we find ourselves drawn toward a different sort of account of the development of a technology. Invention ceases to be an event and becomes instead a process, "a process in the course of which the thing invented changes, coming in time to take forms and serve functions that were no part of the original expectations" (Aitken 1978, 92).

To illustrate this important point, which is fundamental to the new sociology of technology and has important consequences, I want briefly to discuss the invention of the electrocardiograph, the device with which the Dutch physiologist Willem van Einthoven was able to record the electrical activity of the human heart. Van Einthoven received the Nobel Prize in 1924, and the electrocardiograph has become an established instrument in the armamentarium of diagnostic medicine (though unlike Roentgen's invention it did not become the basis for a new profession). How it did so illustrates the way in which some contemporary sociologists analyze technology.

Electrophysiological investigations were already in progress when Einthoven began work. With the aid of a device known as a capillary electrometer A. D. Waller in London had been able to produce the first record of the electrical activity associated with the human heart beat in 1887 (Leibowitz 1970). In 1888 Einthoven saw a demonstration of the use of this instrument and began work along similar lines in Leiden. Einthoven soon saw fundamental limitations in the capillary electrometer and was eventually able to develop a simpler, more sensitive, and more accurate means of measuring the heart's electrical signals. His device employed a silver-plated quartz fiber, tautly suspended between the poles of a powerful electromagnet. The shadow of this "string" was projected by a large arc lamp through a special optical system that photographed the movement of the string shadow,onto a moving photographic plate. The magnet was large and cumbersome and had to be cooled with a water jacket (Barron 1950). In 1903 Einthoven presented the results of his work in a leading physiological journal, *Pflügers Archiv für Gesammte Physiologie.* In 1904 he compared his results with those of Waller and was able convincingly to present his instrument as eminently suitable for cardiographic investigation.

Subsequent events include, significantly, a process by which the electrocardiograph (ECG) acquired clinical significance and, in essence, transformed the meaning of the heart's electrical activity. A second point has been made by Burch, a historian of electrocardiography. When it could only be investigated with Waller's instrument, which had been difficult to employ, the scientific community simply had no interest in the heart's electrical activity. "It was the simple instrument of Einthoven and his clear description of certain electrocardiographic principles that made clinical and theoretical electrocardiography possible" writes Burch (1961), and thereby, he might have added, gave those measurements a new meaning. So far as the establishment of the *clinical* significance of the device is concerned, this is generally seen as having been as much the achievement of the English physician Sir Thomas Lewis, "who transformed the electrocardiograph from an obscure laboratory instrument into a familiar medical tool" (Howell 1984). Lewis had been working on recording and analysis of pulse waves, using a mechanical device called the "polygraph." In 1908 he visited Einthoven in Leiden and returned with an ECG machine to London. Lewis founded a new journal, *Heart,* produced a "massive quantity of original research" and trained "students who were to become the leading cardiologists and electrocardiologists of the next generation" (Howell). Lewis's work with the ECG was largely an analysis of irregularities in the heart beat, or *arrhythmias,* effectively a continuation of his earlier work with the polygraph.

Although Lewis and others were beginning to produce scientifically interesting results, the possible clinical as distinct from laboratory use of the instrument was still a matter of speculation. There
was a major job of education to be done, into which Lewis threw
himself. A lecture he gave to the Royal Society of Medicine in 1911,
"The electrocardiographic method and its relationship to clinical
medicine," shows something of the significance the ECG had at that
time. Having described some of the research results he and others
had obtained, Lewis went on to speculate on the significance of the
instrument and the kind of data it yields:

As part of the equipment of a modern hospital or physiological laboratory
it is already almost a necessity. . . . A few minutes (three or five) places the
clinician in full possession of all the necessary data for the analysis of an
arrhythmia in all but exceptional instances, and the data are of the most
accurate nature. In several respects it presents material advantages over
the polygraphic method . . . the ease with which the majority of the records
are read; polygraphic curves often require minute and often tedious
measurement; the majority of electrocardiograms may be read at first
sight.
    The comparison of one method of physical examination with another
is always difficult, but restricting comparisons to hospital patients, I have
little hesitation in stating that in the routine examination of the heart
patients the galvanometer method affords, or will afford in the near
future, information of equal or greater value than any other single method
at our disposal, be it instrumental or subjective. It must be allotted at least
an equal place with percussion and auscultation, with sphygmomanom-
etry and radiography. (Lewis 1911)

But if Sir Thomas Lewis was convinced that the ECG had almost
earned its place in the hospital ward, his audience saw things
somewhat differently.[9] No ill-trained country healers, the ECG was
nevertheless an arcane device for them, and Lewis's talk proved
hard going. In the discussion recorded in the *Proceedings,* speakers
all refer to their lack of familiarity with the device and their difficulty
in making relevant comments on Lewis's report. The audience was
clearly unsure as to the clinical future of the ECG. In other words,
if our task is to account for the development of the electrocardiograph, which we now know as the indispensable tool of modern
cardiology, we would have to concede that this belief did not exist
in the London of 1911.

Investigators and clinicians were, however, slowly gaining some
familiarity with the new instrument. The Cambridge Instrument
Company, at Einthoven's personal instigation, had begun making

electrocardiographs, with the first device supplied in 1911 and "about twelve outfits" by the outbreak of war (Barron 1950). By the end of the war, Barron argued, "the electrocardiograph became fully established as a clinical instrument for hospital and consultative work." By 1918, the instrument had nearly achieved its modern significance for physicians. Let us look at what cardiologists were doing with the electrocardiograph. Recent historical work suggests that, by this time, Lewis himself was becoming disenchanted with the electrocardiograph—coming in fact to doubt its clinical value (Howell). Partly due to the cumbersome design of the instrument, Lewis's dissatisfaction had more to do with the questions Lewis was posing. For him, argued Howell, the essential clinical question was that of arrhythmias and in particular diagnosis of the condition known as atrial fibrillation. But not only could this condition be diagnosed quite easily using the unaided senses, the therapy available at that time for cardiac arrhythmias involved no possible role for the ECG. The real (i.e., modern) clinical value of the ECG was established not by Lewis in London, argued Howell, but by James Herrick in Chicago.

In 1912, in a paper published in the *Journal of the American Medical Association,* Herrick described the clinical syndrome of myocardial infarction to the English-speaking world. His paper "fell like a dud" as he himself later observed. But in 1918 in experimental ligation of the coronary arteries of dogs, anatomic and electrocardiographic effects were studied simultaneously. Here was a turning point. The concept of the heart attack had arisen. In Burch's words, people now "began to die of myocardial infarction rather than indigestion," for in 1919 Herrick and Smith were able to show clear changes in the ECG pattern associated with "heart attack." This paper attracted much more interest than that of 1912 and is important, Howell argued, for a number of reasons. On the one hand, these electrocardiographic representations of infarction seem to have played a major part in bringing about acceptance of the syndrome itself (implicit in the notion that people should "begin to die of heart attack"). But at the same time (a sleight of hand by which, as so often, history defeats logic) the establishment of this ECG representation of heart disease stimulated an appreciation of the broader possibilities of the instrument in diagnosis. "The electrocardiograph began to serve as the objective graphic method for establishing firmly and convincingly the clinical diagnosis of coronary heart disease" (Burch).

This brief account shows something of the nature of a sociological perspective on technology. If what has to be explained is the "constitution" (or even "existence") of what we *today* call an electrocardiograph, then we should not claim that this took place in 1903 with the publication of Einthoven's first announcement of his galvanometer. Like Aitken, we can see that there are good reasons for preferring to speak of the electrocardiograph being brought into being—I shall use the word "constituted" hereafter when I want to make clear that I intend a sociological perspective—over a prolonged period. It was only with the publication of Herrick and Smith's 1919 paper that the ECG began to acquire the meaning that it essentially has today—that is, as an authoritative arbiter of the presence or absence of organic heart disease. Even then, there are arguments for saying that the apparatus in its modern sense required a still longer period for its constitution, since we should want to be sure that the association with heart disease was not the private intellectual property of an elite group of cardiologists, but that physicians in general, and heart patients too, had come to vest it with this significance. Some might claim that only when the overweight middle-aged man turns in anguish to look at his ECG trace has the instrument really been invented!

These notions, the idea of a technology being representable as a set of distinguishable meanings associated with different connections with or interests in it on the one hand, and the idea of its being constituted through the gradual attachment of these meanings to it on the other, are central to the sociological approach I am presenting here. It is an approach that derives from the application of recent perspectives in the sociology of science to the study of technology. In order to explain in a little more detail what these present sociological perspectives amount to when applied to technology, it is probably simpler to let the authors speak through their own studies.

In much of the work of Latour and Callon the scientist comes clearly and explicitly to occupy center stage: he or she is now the primary social actor, doing constructive work (Latour 1983, 1984; Callon 1985). Moreover, the distinction between Nature (and the scientist's grappling with nature in the laboratory) and Society (with which the scientist must also grapple) is dissolved. The sociologist is obliged to treat these two sorts of grapplings identically: "constructing a scientific (arte)fact" means simultaneously constructing a bit of nature and a bit of society. Scientists seek to enhance their standing, their credibility, by establishing the importance of their own problems, their own expertise. To achieve this, they must form strategic alliances, must persuade significant others to share their own definitions and evaluations. A central term here is "enrollment," the successful achievement, through "multilateral negotiations, trials of strength and tricks" of a situation in which all possible allies have come to share the scientist's characterizations and the identities that go with them. It is the scientist who innovates here, who is uniquely in a position to do the work of transformation, of mobilization (enrollment), which innovation actually is. The scientist's search for credibility, for "indispensability," the enrollment strategies which that requires, leads to forming crucial alliances in the economy, in the community.

This perspective relies for much of its power on its style of writing, but also on a particular notion of "explaining" which rings strange to many Anglo-Saxon social scientists. Explaining here is essentially explaining *how*, not explaining *why*—common enough in historical writing though rare in Anglo-Saxon sociology.[10] But what it certainly does do, is make the essential connection between cognitive and social innovation. Roentgen's invention contained within it both a new way of looking at the human body and the latent role that we now call radiologist—that is, people who would actually use the apparatus to look. To conceive of a new way of investigating the world—in our case a new way of investigating illness or the sick human being—is at the same time to conceive of a new set of social roles and identities.

A related perspective has been proposed by Wiebe Bijker and Trevor Pinch (Bijker 1984, 1986; Pinch and Bijker 1984), who stress the *contingency* of technological development. Theory must above all clarify the micro processes of selection through which some "possibilities" inherent in any technology are singled out, survive, and are further developed whereas others are not and do not. These processes are to be studied in terms of the "relevant social groups" that in a sense "define" a technology through their differing views about what in it is problematic and should be improved. Here too the existence of such groups for which the technology has meaning(s) is essential to its existence. For Bijker and Pinch the "stabilization" of an artefact is the constitution of such groups. Such relevant groups will include, though they certainly need not be limited to, producers and principal users. Again we have the notion of technological change linked in some essential manner to social change and now made dependent on it. To pursue our earlier example of the electrocardiograph, we now have to see the modern instrument in terms of the set of groups (producers, physicians, technicians, patients, and so on) for whom it is of significance. Bijker has developed this perspective through a number of historical case studies, of which his study of the emergence of the modern bicycle is perhaps the best known. We should avoid describing "an" artefact as though its individuality were obvious, Bijker argues; it "is" "a variety of things." The evolution of "the" bicycle in its modern form was the contingent outcome of a variety of social processes in which alternative design conceptions and interests constantly clashed.

Where Callon and Latour's world is that of the hedgehog, knowing one big thing, Bijker and Pinch's technology is the product of a world of foxes.[11] Whereas for Latour and for Callon technological innovation is achieved through the constructive work, the enrollment activities, of one scientist possessed of a vision, for Bijker and Pinch the burden is shared. For them, development is a contingent process rooted in the clash of alternative interests and preferences. Scientists, producers, users, and others can all play a role, and outcomes remain uncertain. Here we do not know who—and therefore what—will win. There are always alternatives, and the historian or sociologist of technology is enjoined always to bear this uncertainty of outcome in mind. Despite these important differences in emphasis these perspectives have in common the crucial notion of innovation—the "social construction" of a technology—as a *process* having both temporal and social dimensions. Yet Bijker and Pinch's perspective will not wholly suffice as a sociology of technology. The social groups that play so important a role in this perspective emerge from thin air. Though their problematizations of the technology derive from their experiences and preferences, we know nothing of the structuring of their experience or how these relate to the technology. Nor is there any sense of the different resources of power or expertise they can deploy in articulating their preferences. Yet surely, to revert to the story with which I began, any (dis)satisfaction with the working of the PET scanner that Sochurek may feel does not have the same significance for its future as does the assessment of the physician in charge of it. In general there is too little sense of the fact that technological change takes place within social structures, which both define certain relations to existing artifacts and make for (unequal) influence over their future development. That is the background to my observation, that the structure of relations that have developed between radiologists and the firms supplying instruments to them is in some sense vital to subsequent innovation processes in diagnostic imaging.

Toward a Problematics of Innovation

Although the actor perspectives developed by sociologists have generated some interest among historians of technology, whose focus is frequently on the heroic molding of nature and society, they have been largely ignored by economists. Economic reasoning suggests features that sociologists have mistakenly tended to ignore. The economist recognizes that, though many groups may have preferences, they do not all have identical possibilities of expressing these preferences (for example through purchasing power). Economists argue that the behaviors of buyers and sellers are interdependent, with their transactions, and mutual adjustments taking place in a "market." Industrial innovation can then be seen as an example of "seller behavior," and it too is necessarily affected by market structure. Most simply, the behavior of buyers as expressed in the market determines what can be sold and thereby constrains the innovation process. Of course in real economic theorizing things are much more complicated, and for the economist the *variety* of innovations is a function of the variety of markets. Hence the interest in variables such as concentration ratios, merger behavior, and entry barriers that capture differences in market structure. A theory that fails to conceptualize market as essentially constituting the relations between buyers and sellers, as providing the locus for their transactions, is thus irremediably flawed from the economist's point of view.

While this lack both of any notion of structure, of any locus for transactions, is indeed a fatal flaw in much recent sociology of technology, sociology similarly points up a lacuna in much economic thought.[12] In asking what does structure economic behavior, and industrial (product) innovation as a special sort of economic behavior, the sociologist is not limited in his or her analysis to *market* relationships (even if extended to include government as a regulator of markets). The sociologist can allow for more complex and diverse relationships to the innovation process. For example, in certain areas in which innovation takes place the relations between buyers and sellers might be very different indeed from the usual economic one (see for example Fries 1988) . This very point was made in memorable and highly pertinent fashion some thirty years ago when President Eisenhower offered the notion of a "military-industrial complex." In warning us of its dangers, Eisenhower was alerting us precisely to the fact that in such a "complex" *shared* interests between buyers and sellers can override the opposition of interests implied by a price mechanism. It suggests a common interest in increasing the size of the market and the rate of innovation and in preserving the autonomy of the market from external control or regulation. In Eisenhower's example this common cause finds expression in the political system. Something similar, though otherwise expressed, characterizes the process of innovation in diagnostic imaging, as we shall see. Let us first look more generally at how the need for a perspective that seems to draw both on sociology and on economics arises.

To try to join together bits and pieces taken from economics with bits and pieces taken from sociology, if motivated by no more than a spirit of social scientific ecumenism, would not be a very worthwhile endeavor. There is far more at stake than that. To try to come to grips with the real problems of control (as well as the opportunities) that technology poses is a major challenge to the social sciences today. Technology-in-society clearly transcends, both in its complexity and in the problems of social control it raises, the theoretical specifications of the individual social science disciplines. So far as medical technology is concerned, we saw in chapter 1 how fundamentally implicated it is in processes of professional control and domination and in interprofessional rivalries, and how significantly it affects the organization of medical care. We saw something of the complex of interests both social and economic that underpin the innovation process.

Historical facts are easily reconciled with a variety of perspectives. Consider, for example, the modifications to the x-ray tube introduced in the period after 1918 and described by Barclay . We saw at that point how it would be possible to make sense of the modifications that were actually made in terms of "a characteristically iterative process, in which *both* demand and supply side forces are responded to." Nathan Rosenberg would have us make sense of these developments through seeing how manufacturers of x-ray tubes were endlessly sensitive to their own wishes to reduce production costs, to the preferences of their customers, and at the same time to the technical possibilities being thrown up by their own and others' research. Remember that science, research, and knowledge are important variables in his approach. Against Schmookler, Rosenberg argued the importance of the (advancing) supply of scientific knowledge in setting the margins of the possible. Research is the essential means by which entrepreneurs can seek to actualize opportunities in the market or to reduce production costs. Bijker and Pinch could also make sense of the data provided by Barclay. They would stress that these modifications taken together are the outcome of a complex social process in which the various interests in the functioning of the primitive x-ray tube came together and were resolved into a development process. If we were to look carefully, in their manner, we would find that those early radiologists had certain sorts of worries. (For example, as Barclay also tells us, the x-ray tube was unreliable: it required a good deal of craft skill to make it not simply go purple. Perhaps some were also beginning to worry about safety—both their own and their patients'.) Doubtless the manufacturers also saw possibilities, as did makers of photographic film and electrical installations and medical administrators, among others. Some must have lost out; perhaps makers of photographic glass plates tried in vain to improve their product. These accounts tell us different things. Bijker and Pinch remind us that, undoubtedly, not all proposed modifications were taken up; that some interests in the further development of the x-ray tube proved decisive; that we must look more widely at the full range of interests and preferences to make sense of what *actually* emerged. With Rosenberg, our account is more closely geared to the manufacturer's point of view. The manufacturer takes note of the preferences of his customers in proportion to their significance—which may be their purchasing power. It is not just that these analyses are complementary and we need both. It is rather that to make sense of the salient issues that imaging technology raises we need to try to understand the expression of preferences with which Bijker and Pinch deal within a structured locus of exchange such as the economist's notion of a market represents. How can we theoretically capture what happened as the x-ray industry gradually tailored its innovation to the expressed needs of

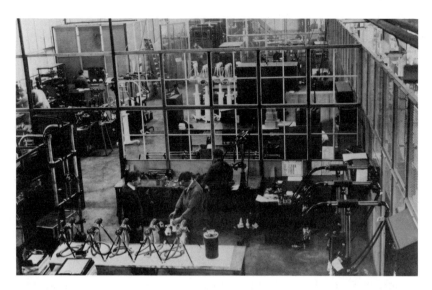

Figure 2.1
Manufacture of x-ray equipment, c. 1932. Completed tubes, similar to those shown in figures 1.3 and 1.5, can be seen in the foreground. Reprinted by permission of Philips Medical Systems.

an increasingly professionalized group of users (see figure 2.1)? How is the innovation process affected by the seeming importance of new medical technology for professional and institutional status?

We seem to need a theoretical perspective in which technical and economic activity are embedded in, and somehow interact with, a more complex social structure. Innovative activity on the part of a firm is "boundedly rational," but what bounds the rationality is no longer only lack of perfect knowledge. Rationality is also bounded by the "embeddedness" of economic activity in social life. We saw in chapter 1 how GE decided to manufacture Coolidge's hot cathode x-ray tube despite their unfavorable market estimate, and for reasons undoubtedly to do with a broader assessment of the company's interests. According to Granovetter, criticizing Williamson, markets are typically marked by social connections having nothing to do with economic rationality (Granovetter 1985). Real buyers and sellers often become friends, and directors of competing firms may belong to the same clubs, sit on the same committees. In Granovetter's perspective, firms appear as "regions" within broad networks of social relationships linking firm with firm and criss-crossing markets. These relationships will frequently set bounds to rational action as conceived by the economist. The assumption of rational action on the part of economic actors should be the starting point—"a good working hypothesis that should not easily be abandoned"—but its real limits should be empirically examined in sociological accounting for economic behavior.

Where Granovetter's network of social relationships derives from the attempt sociologically to account for economic activity in general, other social scientists are moving toward a rather similar position in the attempt specifically to account for technological innovation. Wesley Shrum has focused on the growth of technical systems, which he defines as "centrally administered network(s) of actors (organizational as well as individual) oriented toward the achievement of a set of related technological objectives" (Shrum 1985, 15). The interorganizational character of the network, with only partly coupled units having only partly consonant objectives, is central to Shrum's perspective. Development of a technology (nuclear waste disposal and photovoltaic cells were studied comparatively) is then to be understood in terms of the structure and functioning of the appropriate network.

Other authors use somewhat different concepts. Both Burns (1985) and Law and Callon (1987) emphasize innovators' mobilization of resources of social, economic, or political power. Frequently, the resources needed to bring about the social reorganization entailed in technological change exceed those of any individual participant. Burns refers to the importance of coalition building, involving diverse participants "linked across boundaries of specialization and institutional spheres by means of market networks, business and ownership structures, through industrial and public organizations, as well as through research and technical networks." Law and Callon distinguish between what they term "global" and "local" networks. The global network is used to produce the resources necessary to sustain a local network within which the project will be brought to fruition. Law and Callon relate their network concepts to economic notions of "product differentiation" and "barriers to entry." Thus, the attempts made by actors to "monopolize a local network" are equivalent to the erection of barriers to entry to a given market, while differentiating one's product from potentially competing products amounts to monopolizing the global network. Here we see a hint of how structural notions might constrain the form given to a product emerging, for example, by indicating the need for substantial product differentiation. Burns, like Bijker and Pinch, emphasizes the contingent character of the precise embodiment of the technology that emerges: "In general, technologies—and the sociotechnical systems in which they become integral parts—may take a variety of forms and properties. Which one will emerge and be successful—or which of several equally promising forms—cannot be determined a priori. This reflects the fact that technology developments . . . entail a variety of technical, economic and sociopolitical problems which must be solved in concrete settings by a constellation of actors." The notion that successful innovation entails restructuring social relationships is one that we have already encountered. We saw in chapter 1 how the constitution of x ray as a diagnostic device entailed adaptation in both medical and industrial structures. Burns speaks also of the coevolution of technology and social structure through a process of mutual shaping: "New industries— their production processes, the goods and services they produce, and their distribution and utilization—make up complex sociotechnical systems. The 'new' has to grow within the framework of the 'old.' It cannot be clearly perceived. The new is created based on experiences and innovations with the 'old,' the established sociotechnical systems and institutions. Social actors—private and public, individual and collective—play the roles of entrepreneurs and change agents. . . . They bring about—or try to bring about— organizational and institutional changes which will facilitate or validate the new products and systems" (Burns 1985).

Abernathy and Clark take a step further in suggesting that innovations have to be disaggregated—that while some innovations might entail major social restructuring, many do not (Abernathy and Clark, 1985). Simplified, Abernathy and Clark's scheme involves a contrast between "regular innovation" ("change that builds on established technical and production competence and that is applied to existing markets and customers" and the effect of which is "to entrench existing skills and resources") and "architectural innovation" (which uses "new concepts in technology to forge new market linkages" and may often lead to establishment of a new industry). Abernathy and Clark's own illustrative material relating to the history of the auto industry as well as detailed studies of industries (e.g., Dorfman 1987) and of firms by business historians (e.g., Graham 1986) are beginning to show the diversity of innovation processes found in a single industrial sector. The outcome of the innovation process is highly uncertain, not only because of the recalcitrance of nature and the imperfect information of the innovators, but also because conflict with vested interests can arise. The success of a technology depends either on the ease with which it can be accommodated in existing structures of relationships or on the ability of its proponents to mobilize resources needed for restructuring (e.g., the establishment of a new industry). If they fail, then the new technology will either be totally blocked or, possibly, marginalized. If the proponents of a new medical technology fail to secure its integration into the appropriate structures of professional practice and industrial production then, according to Burns's view, it will fail. So far as diagnostic imaging technologies are concerned, this implies that success should depend upon compatibility with, and successful integration into, the interorganizational structure in which radiology and the x-ray industry collaborate.

Fundamentals of the Present Analysis

Imagine an industrial scientist with an idea for a radically new product approaching her laboratory director and, through the director, management. The idea has been suggested by her research, which is of a long-term exploratory kind, and the scientist is excited and enthusiastic. She wants to go further, but to do so she has to gain support. Management, she knows, looks at it differently. The inherent significance of the research is of minor interest, at best, and will not of itself win support. Discussion is going to hinge on costs and benefits—on preliminary assessments of what it will cost, in cash and opportunity, and the benefits any such product might yield. Managers not only speak another language from the scientist, they have other commitments. Rough estimates suggest that costs, a major consideration, could be very high. Is there any chance of obtaining a subsidy from a government innovation program? And what if the scientist is correct, and a product of the kind she is suggesting could be developed? What are the potential benefits? What would the market be, both qualitatively and, at a rough guess, quantitatively? Much of the market may lie in a sector with which the firm has little experience. Would it be able to compete there, however promising its product?

It is not too difficult to imagine this as the bare bones of the first phase of the development of a new medical technology. If we were to write this as a historical case study, given the ideas that have been sketched out, there are certain characteristics that it would have to have. In the first place we would want to write it in such a way as to disclose the characteristic interests or commitments expressed and articulated in the interaction between scientists and management. To do this, we require some sense of what these commitments are. We would want to take the actors in our accounts as in some sense articulating the collective commitments of scientists or of managers/industrialists. The sociology of science has a lot to tell us about the collective commitments that underpin scientists' actions. Robert Merton showed decades ago that something of the way science works could be understood in terms of scientists' search for the approval of their peers through their contributions to the collective endeavor. From the work of Merton and his students we learned that publication is vital to the scientist because it is the public expression of a claim to have discovered something significant (see Merton 1973). Through the approval of his peers the scientist acquires not only public esteem and recognition but also the resources needed to further develop his work and to advance his reputation. Similarly, of course, it is possible to speculate about the collective interests on which entrepreneurial or managerial behavior rests. This is a matter on which there is no shortage of theorizing, and Schumpeter's Promethean entrepreneur has little beyond the history of economic thought in common with the colorless peruser of balance sheets of much later writing. Again for our present purposes we need no more than a very general notion, and whether we take growth of the organization (firm), or profit maximization, or some complex function linking the two is of no consequence. A general notion of management as the (boundedly rational) search for profitable expansion of business is adequate for the moment.

Developing our account of the mythical scientist's confrontation with management, and what happens next, places more demands on theory than simply indications as to the interests expressed. Such a confrontation is more than the clash of interests—the scientist arguing the interests of science, the manager those of business—because it demands an intellectual synthesis. How is a compromise established between a sense of the technological possibilities suggested by the new research and the complex commercial considerations deriving from business experience? All is surrounded by uncertainty; no one can be sure that the scientist's hunch will be right, or that the assessments of costs and markets are even of the right order. This is the message of Nelson and Winter's behavioralistic approach. Recall how they speak of "backing and forthing" in the formulation of an R&D strategy. They also tell us that firms rationally try to reduce the uncertainties involved in making decisions and to reduce the risks. How might they do that here? Seeking government subsidy for the development work is one way of reducing the risks. Taking over a firm with well-established links into the medical devices market is one means of uncertainty reduction. Building up ("enrolling") support in the way the scientist tried to do in approaching management is another element in the strategy behind the development—or constitution—of a new technology. In writing the histories of the technological developments, we must characterize the actions of the principal actors both in terms of the objectives they seek to attain and of the strategies that they adopt.

If we extend our story, we can imagine the firm successfully securing a subsidy for the further development of the new technology. Work will go on. The firm knows that markets work in very different ways, and that one of the unique features of the medical instruments market is that competition takes place less on the basis of price than of performance. Adapting its emerging product to the criteria and preferences of the new market requires intelligence based on firsthand knowledge of the needs of the physician. What do they expect from a high-performance instrument? We can imagine the firm as having organized an appropriate input into its further development work. This will now take place in partnership with an R&D organization on the market side—something like the "quasi-integrated forms in between pure forms of internal organization and market organization" to which Granstrand referred. In the case of a medical technology this might be a university hospital. Further historical analysis would then focus on the collaborative research and development work—on the choices made, on the evolving sense of the precise place in medical practice to be aimed at in further development.

As work proceeds, as the first working prototype more than fulfills expectations, new demands emerge and the strategy of the collaborating scientists and engineers changes. Management has to be kept interested and may demand more proof of market possibilities. At the same time the researchers typically begin to publish some of their work in scientific journals. To understand what and why they publish and the significance of publications for our reconstruction of the innovation, we have to look at more recent refinements of Merton's model of the scientific system.

The idea of a relationship between publication and reputation remains, though sociologists of science now think in terms of reputations largely made within specialized professional communities (which Whitley (1984) calls "reputational organizations"). A scientist derives his or her reputation from the acceptance of successive "knowledge claims" by peers. In other words, a physicist's reputation will be advanced by publishing studies of appropriate problems, using the methods generally employed by physicists and in journals read by other physicists. Reputational "profit" depends upon how skillful, how relevant, how fruitful the work is judged to be. As they develop and improve their technique, physicists thus have a reputational interest in rationalizing and reflecting on the steps through which they have worked. Medical scientists have to publish different things to earn the respect of their peers: unraveling the etiology of a disease, showing how it can be more rapidly or more effectively diagnosed or treated, and so on.

We know now that publications have a contingent relationship with past laboratory work. Scientists publish neither everything they did nor precisely what they did (Holmes 1984). Publications focus on what is felt to be significant for peers, and they try to establish value by building connections with other significant work. It is also becoming clear that empirical work in the laboratory does not in itself "determine" the knowledge claim that becomes attached to it. Making use of theoretical, logical, and stylistic resources scientists can present their work in different ways for different audiences (Amsterdamska forthcoming, Myers 1985, Pinch 1985).

The view that reputations are made within structures (specialties or disciplines) having a degree of coherence and continuity, which at the same time reward past achievement with the resources needed for further work, has also been challenged. Knorr-Cetina, for example, has argued the importance of contingent and local "transepistemic arenas" as the extended social locus of scientific work (Knorr-Cetina 1981, 1982). Not the speciality but a heterogeneous grouping of diverse interests assembled by the scientist provides the locus for the decisions and negotiations on which laboratory work depends. Other scholars have argued that the functions of scientific disciplines have to a growing extent been usurped by "hybrid communities" (van den Daele et al. 1977). Composed of scientists from more than one discipline, as well as nonscientists, and organized around external problems deriving from science policy programs, it is these communities that are said to provide resources and reputations.

Our imaginary example involves physicists or engineers working together with medical scientists. Out of the interwoven texture of their work are drawn threads of physics, threads of medicine, to be presented to appropriate professional audiences. Publication serves a dual purpose. One, following obviously from what has just been said, is the advancement of scientific reputation. The second reason is more complex. It depends upon a notion of how the publications are read. Ideally, leading figures in the medical profession, reading the papers, become intrigued by the possibilities of the new device. In that way a receptive climate is created in advance of commercial availability. Following Knorr-Cetina, we see how this can be used to bolster claims for support for the project within the participating organizations. It bears witness to potential market interest. What might also happen is that other groups, reading the papers, decide to replicate the work. The more promising the device, the more available the kind of knowledge on which its functioning rests, the more it seems likely that others will be spurred to try to build a similar device (Collins 1974). Competition may be purely scientific, or it may (soon) involve competing industrial interests. The seeds of future commercial competition may have been planted.

As our account of the machine develops, we need to try to understand how the strategies of the organizations involved are shaped and how they evolve. It is through these strategies and the choices they entail that the scientist's original "vision" is realized. As the device approaches the marketing stage, new considerations begin to weigh more heavily.

What sort of a market might our imaginary firm confront, and how accessible might that market look? How is the perceived structure of the market likely to affect the strategy, the behavior, of the firm seeking access for its new product? What of potentially competing products based on replication of the initial research? To continue on we need to introduce a few of the structural concepts used by economists. What sort of entry barriers might our firm confront? How likely are existing producers to be able to keep hold of their markets? Should the product to be sold as competing with traditional ways of doing things (for example as a replacement for x-ray technology), or should complementary functions and uses be stressed?

Under certain circumstances purchasers tend to stay loyal to their traditional suppliers. On the basis of a study of various segments of the computer and semiconductor industries, Dorfman argues that

IBM's remarkable success in holding on to its mainframe computer market is partly attributable to a number of aspects of the firm's strategy (Dorfman 1987). One is the level of its investment in R&D, which has enabled it not only to produce a stream of new and improved products but also to retain the all-important image of a firm at the cutting edge. No less important, argues Dorfman, is the fact that mainframe computers were marketed as components of complete systems, thereby requiring a major commitment on the part of the purchaser (including peripherals and, above all, software). At the same time the firm offered a variety of services, including maintenance, education of personnel, and software support, which in a sense build up a dependence. Under these kinds of circumstances, and especially when the systemic nature of the total package makes comparison of alternatives difficult, purchasers tend to want to play safe and stick to known and trusted suppliers. Companies wanting to enter such a market were then well advised to seek not to compete but to look for specialized niches (e.g., supercomputers). Other segments of the market, according to Dorfman's analysis, prove to be quite different. An example is the market for magnetic disk storage, which is very different partly because of the characteristics of the technology. Disk drives differ in respect of very many characteristics (e.g., storage capacity, access speed, compactness, reliability, interchangeability, and, crucially, price). Because of these differences, all of which are relevant to the customer, there is scope for a variety of tradeoffs. Different combinations of characteristics will suit the needs of different classes of customer, and there is room for avoiding competition on this basis. Dorfman writes, "The more technically complex the product, the less technically sophisticated the buyer, the harder it is for the customer to experiment with competing brands, and the less frequently the customer purchases the product the greater is likely to be the tendency to choose a product initially on the basis of the image of the product or the firm, and to stick with the initial choice, even at some sacrifice in price, so long as the product is satisfactory" (Dorfman 1987, 238). This loyalty seems to be as typical of hospitals in regard to their purchase of major equipment (Hamilton 1982, 79–80) as it does of purchasers of mainframe computers.

Sometimes existing firms do fail to take advantage of a new technical opportunity, and the newcomers move in. Of the firms making vacuum tubes, for example, almost all (with the exception of AT&T) miscalculated the potential of the new transistor technology. As early as 1953 the major U.S. vacuum tube manufacturers (which included GE, RCA, and Raytheon) started to produce transistors, as did a number of newcomers (Texas Instruments being one). By 1963, the newcomers had claimed 68 percent of the market (Dorfman, 177). Why did the vacuum tube companies fail? Dorfman suggests that among the reasons were their failure to see the transistor as anything more than a smaller substitute for the valve; the inability of firms with established technical staff in senior posts to compete for scarce new expertise; and the major organizational changes in engineering and research departments necessitated by a totally different technology, and which firms were unwilling to make (Dorfman 1987, 217–218).

With concepts of this sort we can begin to appreciate the situation confronting a firm looking to enter the diagnostic imaging market with a radically new product. Serious concern for its potential market implies attention to how precisely the new product is seen as fitting into the existing structure of medical practice. What will it do? The question of whether to offer an alternative to conventional x-ray or a technique that complements existing diagnostic practices is not simply a matter of marketing. The ground will in a sense have been prepared by the medical participants in the enterprise. Their research, using the device first on volunteers and then on a small number of selected patients, will have become central to further development. Through this work, some sense will have been established of the kinds of utility that the new device might have. Suggestions to this effect will appear in publications from the project, though consensus will be reached only slowly. Research continues to play an important role in establishing the utility and efficacy of the new device. Now of course it is not restricted to the initial researchers, not least because there are general reputational interests in posing questions relating to the use of a new device. Russell's work on the diffusion of new medical technologies suggested that to earn its place in medical practice, the new device may first have to become a recognized focus for worthwhile and interesting investigation.

Our mythical account has now served its purpose, and we can leave it there, with clinical research in full swing! We must now try to extract its central elements in order to establish a structure for comparative analysis of actual imaging technologies. The account is constituted around three concepts: *interorganizational structure*, *career*, and *problematization*. I want now to try to clarify these concepts.

The notion of interorganizational structure has been presented earlier. It captures the fact that we are concerned with diagnostic imaging devices and not airplanes or video recorders through characterizing the specific structural relations between producers and users. I have hinted at various signs of the interweaving of market aspects of the supply of x-ray equipment with the practices of radiology. Individual hospitals, as consumers of sophisticated imaging equipment, seem as bound to their particular suppliers as (according to Dorfman) mainframe purchasers were largely bound to IBM. This is achieved and maintained in all sorts of ways. Postsales contact—the servicing of hospital equipment—is a matter which may have slipped beneath the attention of most students of innovation but which may contribute importantly to this kind of nonmarket integration. Cockburn's account of the arrangements made for servicing notes how x-ray equipment is typically designed "so that whole components are removed and replaced when faulty. . . . It is almost impossible today for a hospital to avoid signing contracts with the suppliers of their equipment, who send their own engineers to the site. The manufacturers engage in a similar struggle with the [hospital physicists]. 'They want to talk with the computer and play with it,' complained one company's technical manager. 'We don't want that. We've designed it for simplicity in use. They take too intelligent an interest' (Cockburn 1985, 134).

Information flows between imaging department and industry, in the interorganizational structure (or organizational field) which they constitute. In line with DiMaggio and Powell's perspective, professions such as biomedical engineering and medical physics integrate the two (DiMaggio and Powell 1983). Competition between hospitals is based on status rather than on the efficient use of resources, and possession of the most advanced technology is a crucial aspect of status. We saw earlier how little inclined leading hospitals were to await the results of careful evaluation before purchasing advanced technology. The whole field is characterized by shared commitment to technological innovation, to continuous improvement to a technological armamentarium vital to the status of both industry and hospital. Yet these interests might also set limits to innovation. Radical innovation, argue Abernathy and Clark, "disrupts and destroys. It changes the technology of process or product in a way that imposes requirements that the existing resources, skills, and knowledge satisfy poorly or not at all. The effect is thus to reduce the value of existing competence" (Abernathy and Clark 1985). Too radical innovation, in other words, calls existing professional expertise into question, while at the same time threatening existing markets.

The notion of *career* plays an essential expository role. Career is defined here in terms of certain conventional phases that—like life-cycle phases such as teenager or pensioner—are intended to facilitate comparison and generalization. It is perhaps worth making clear that these phases differ considerably in the complexity of what they seek to represent from the notion of product cycle used in the economics of innovation. A pensioner is one who has a particular position in relation to the labor market, yet the concept conveys much more regarding other social statuses and way of life. So here too. The phases of the technological career are intended simultaneously to express position in contexts both of production and use. For that reason the concept of career as I mean it here differs from research-development-production models or from McKinlay's somewhat ironic account of the career of a medical innovation (McKinlay 1981). It comes close to Hughes's (1985) invention-development-innovation-momentum model of technology.

It is not that there is a development process, terminated by market launch. Nor is it the case that a medical device starts its life on the ward. Rather the locus in which the career is made is within an *inter*organizational structure. Within this the locus of central constitutive activity shifts gradually from laboratory to industry, to hospital. In each place, at each phase, the new imaging device is shaped and reshaped by those who have become committed to it. The phases are defined as follows.

Phase I, Exploration: The exploratory phase is the history of the attempt at realizing a certain vision—that a particular technological principle might be applicable in the medical area. For an imaging technology it is the period leading up to the first working prototype and the first publications reporting its successful use, principally the research going into the demonstration of its use. In other words, by the end of the exploratory period there should be a plausible demonstration that the principles of image forming do work. The medical and industrial communities are just becoming aware that one or more research groups have achieved an original imaging device.

Phase II, Development: The development phase of an imaging technology is defined as beginning with the publication of the first human in vivo images of the desired type using the first prototype. It ends with commercial manufacture and the new device being brought to market. This phase typically includes significant industrial development work involving the production of a number of prototypes. Whereas in characterizing the four technologies in their exploratory phases we will need to look principally at who was doing the research, the focus of accounting for the development phases shifts to industry. The question becomes, what industry? The focus of analysis here will be on the growth and nature of industrial involvement in the emerging technology—on industrial R&D.

Phase III, Diffusion, evaluation, and assessment: This phase essentially corresponds to the integration of the new technology into medical (radiological) practice. It begins with the first attempt at publicly specifying clinical uses for the technique on the basis of clinical experience and ends with the technique becoming institutionalized in practice, with at least some routine uses attached to it. This does not necessarily imply 100 percent diffusion (with all hospitals required to possess the instrument). It implies a widespread consensus as to good practice—over conditions for routine use. By this stage major decisions will have been made. Firms that have successfully completed the development process will have had to decide to what extent to invest in commercial manufacture. At this stage, too, some response on the part of the existing supply industry (if not hitherto involved) is likely. Whether on the basis of clinical results, or by copying others, hospitals will be deciding whether to purchase the device. Its eventual incorporation into practice implies that certain accommodations are to be made, affecting both the use of other techniques and the organization of work in the hospital. This will entail what Barley has shown (in the case of CT scanning) to be a renegotiation of the hospital's social order—the relations between radiologists, radiographers, technicians, referring specialists, and so on (Barley 1986). In this part of each case study we shall look principally at the kinds of questions asked regarding the new device, the clinical research through which its proper place in the diagnostic armamentarium is established.

Phase IV, Feedback: The feedback phase leads to the development of improved models and the search for new applications if the technique is judged successful. This final phase is initiated by attempts at establishing how commercially available models might be improved on the basis of experience with their use. Such attempts might be made either by manufacturers or by users. So defined, feedback might begin at any point after (and in fact even shortly before) commercialization. In the case of a successful innovation, economic theorizing suggests that feedback presages a new round of development and diffusion as the new model replaces the old. At the same time experience suggests that feedback *can* begin so early that assessment can hardly keep up: the "moving target" phenomenon.[13] In the case of a successful technology, also, feedback should be long lasting, ultimately terminated only by withering either of medical or industrial interest brought about by the rise of a new technology.

The question of a means for comparing innovations in diagnostic imaging naturally follows. In the following chapters, as we go on to look at the "constitution" of diagnostic ultrasound, at infrared thermography, at CT scanning, and then at magnetic resonance imaging, we shall want to compare and contrast their "careers." In other words, we shall want to know how *how* and *why* the initial "vision"— what Mikael Hård calls the "ideal type"—became the precise instrument—the "type"—which actually came to be used in the hospital.[14] We shall need phase by phase to understand how each device was constituted by the physicians and engineers involved. What did they have to accomplish, and how was the work of constitution structured? This is where the third concept, "problematizations," comes in.

The interorganizational structure binding roentgenography and industry both enables and constrains. A sort of "hybrid community," it provides social financial and intellectual resources—a means of assessing what has been achieved, access to the patients assembled by radiologists, and access to the manufacturing and marketing skills assembled by their suppliers. It also constrains. Position in the interorganizational structure is of major significance for the innovation process and influences what has to be achieved. Reflexive participants are aware of this. Here is the founder of the firm Sonicaid, which entered the medical area with a low-priced fetal heart detector in 1967 and has since become a major producer of fetal monitoring equipment: "When we brought out the fetal heart detector, the fetal monitor, in the early days, we were persuading the doctors. . . . Now, as the company has become much bigger one is reflecting much more on what the doctors are wanting. As *you get bound up with the medical profession* development begins to stem much more from feedback from doctors"[15] (Emphasis added).

A firm with long experience in the diagnostic imaging market has a strong sense of what the market wants. Experience, contacts, and knowledge there confers a more rigorous and binding sense of what a new imaging device ought to be like than a firm new to the business has. Insofar as success does depend upon the compatibility of the new device with existing practices—upon success in "engineering" this compatibility—then experience counts for a good deal. Not only is the name of an established supplier trusted, but, it seems, an established firm has a better sense of what practitioners want. If compatibility with existing structures and practices is a sine qua non of success, *then* we ought to find the same structural constraints acting in all cases. The notion of "problematization" focuses our attention upon the "work" of innovation and its structuring.

Problematization refers to what is entailed in some aspect of (an emergent) technology being taken to be problematic. Problematizations can take many forms. For example the establishment of new regulatory requirements in regard to a device, or regulations governing the circumstances under which it can be used or its use reimbursed are instances of what I have in mind. In practice, we will look principally at the research through which the technology takes shape: the questions asked, the terms in which they are posed, the resources brought to bear in attempting to deal with them. I indicated the significance of published research for the constitution of a new medical technology. In the mythical account we saw the research gradually shifting. As the locus shifted from laboratory to industry to hospital, the questions being asked changed from questions of physics and design to questions of safety and clinical utility. Research has a double significance, for while it (partially) constitutes the technology, it is also (partially) constituted by the technology. Looking at what a prototype device will do in the clinic contributes to the process by which that device is made functional, while (in concert with disciplinary concerns) the device shapes the research. I shall suggest that the interorganizational structure centered on x-ray technology defines a series of problematizations. Radiologists will typically compare a new device with their existing practices, judging its performance in terms of established parameters of accuracy and in relation to existing priorities.[16] These matters are of course known to the industry supplying such devices. I shall show that new imaging technologies do not, in fact, always display the same pattern of problematizations. Looked at in terms of problematizations, careers differ, displaying a wide divergence in terms of intellectual resources deployed, publication patterns, and attempts at replication. Radical innovation, threatening established interests, is possible. In such cases the questions asked, the nature of the necessary evidence that has to be collected, and criteria of judgment, depend upon an emergent sense of the possible uses of the technique that may only gradually become apparent. What sorts of responses do such challenges to professional and industrial interests evince? Can we trace this, too, at the level of problematizations—in conflict over priority, methods, and demonstrations of clinical utility?

The accumulation of successive innovations in the techniques of diagnostic imaging gradually bring about structural changes in radiological practice. Other authors have suggested that with the growth and maturation of an industry innovation patterns typifying it gradually change (Abernathy and Clark 1985). The "contexts" into which successive innovations are introduced are not strictly speaking the same. At the same time the broader social, industrial, and demographic structure of society also changes, so that the historian is properly enjoined to embed the history of technology in a more general social and economic context. Historically the periods covered by the case studies that follow—from the beginnings of exploratory work on ultrasound (in the late 1930s) to the market approval of magnetic resonance imaging (in the mid 1980s)—was one in which vast changes took place. World War II intervened with its unprecedented mobilization of scientific talent. In large numbers biologists acquired experience of applied physics and of mathematical techniques as they were set to work on problems of winning the war. This expertise they took back with them to their laboratories, and they brought about a transformation of biology. By the 1950s, with the Cold War in full swing, atomic and space research were becoming major national priorities. Tremendous advances in electronics, in materials science, and in thermodynamics followed. All this was of great consequence for the development of medical technology. Not only did entrepreneurs begin to see medicine as a "fertile field for exploitation," but the medical profession began to develop a growing enthusiasm for new technology (Eden 1984).

These historically specific dependencies were complex. The postwar demobilization of expertise was to prove of immediate importance for the emergence of diagnostic ultrasound. The development of wartime sonar and radar was to have both analogical and practical implications here. Not only was there widespread appreciation of these new modes of spatial localization (known to have been so important to the war effort), but there were also skilled engineers looking for new applications for their knowledge. Firms, too, had spare capacity, and there was undoubtedly an interest in turning swords into the proverbial ploughshares. A few years later medical thermography came to depend on the technologies of the ensuing Cold War. Still later (the early 1970s) we had the development and commercialization of microelectronics and of computers, with great consequences not only for the speed with which technical problems could be handled, but also for the kinds of things that could be done. For example, the reconstruction of images from projections had to remain a mathematical theorem with little relevance outside the laboratory until the advent of cheap computing power. The CT scanning story shows a clear interdependence with the development of the computer and the computer industry at the end of the 1960s. This, combined with the concerns about the safety of x-ray that also surfaced then provide unique temporal features of the development of CT scanning.

History places limits on comparison and sneers at the poverty of theorizing. The development of microelectronics, the new attention paid by governments to stimulating industrial innovation from the late 1970s (combined with the rapidly escalating costs of development work in the medical devices area), and the growth of regulation of the drugs and devices that can be offered for medical use provide a broader context of evolution in time and variation in place than can be handled in the following analysis. I am all too aware of the simplification entailed in comparing developments that took place in such a changed world.

3

The Constitution of Diagnostic Ultrasound

Introduction

Although diagnostic ultrasound has never had the glamor of some of Sochurek's "uncanny new eyes," it has nevertheless had a major impact on the practices of diagnostic imaging. Widely used in modern medicine, the technique's use in obstetrics is most generally known outside the medical profession. For hundreds of thousands of pregnant women an ultrasonic "photograph" of their unborn baby (figure 3.1) is coming to be a valued and unique object. Yet the process by which ultrasonic imaging was "constituted" and achieved the place in medical practice that it now holds is a complex one. Development of ultrasonic technology had been going on for twenty years, principally for military purposes, before the possibility of its application in diagnostic medicine arose. When eventually the notion of diagnostic ultrasound did become a matter for systematic investigation, it had much to do with the "demobilization" of expertise that took place in the aftermath of the Second World War.

The complexities we shall try to characterize in this chapter are not only a matter of the diffuse origins of ultrasonics. They are no less to do with the *process* of its constitution as a medical imaging technology. In terms of the conceptual scheme of chapter 2 ultrasound is a difficult case, for despite their obvious interest in securing control of ultrasound, and hence in maintaining their monopoly over diagnostic imaging techniques, radiologists were not able to do so. The "career" this chapter discloses is a complex one, since diagnostic ultrasound was not incorporated in any simple and unambiguous fashion into medical practice. We will look at how this came about, both in terms of processes at the structural level (formation of coalitions and of markets) and at the

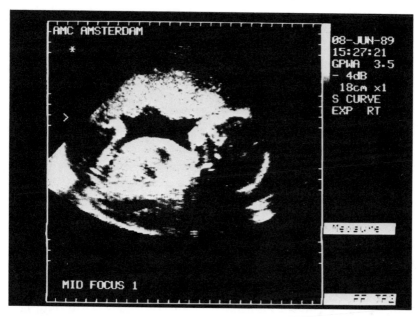

Figure 3.1
Miss Mata Kranakis: The Echograph as cultural good. Reprinted by permission
of Eda Kranakis.

level of "problematizations." To what extent can we make sense of
the career of diagnostic ultrasound in terms of divergences in
research agendas, processes of replication, and demonstrations of
clinical utility?

The term *ultrasound* refers to sound waves—thus mechanical
vibrations—beyond the range of human hearing. The dog whistle
is a well-known example. Unlike electromagnetic radiation (including light and x-rays) sound waves, and ultrasound waves, are
propagated only through *matter*. The composition of a substance,
both at the molecular and at the gross levels, determines the velocity
with which (ultra)sound waves pass through it. That means that
when these waves meet a boundary between two substances, there
is a change in velocity, and at the same time some of the sound
energy is reflected back. These basic properties were described by
Lord Rayleigh in his classic work on *The Theory of Sound* published
in 1877. One method for the generation of ultrasound waves
resulted from the discovery of a curious property of certain crystal-
line materials by Marie and Pierre Curie in 1880. They found that
if a quartz crystal is subjected to an electric field it becomes

deformed, and its mechanical deformation is transmitted into the surrounding medium in the form of an ultrasonic wave. This phenomenon came to be called *piezoelectricity*. Piezoelectric crystals could function both as generators and detectors of ultrasonic waves, since the phenomenon worked both ways.

The whole effect was worked out in much more detail by the great French physicist Paul Langevin, under the stimulus of the First World War. In France, as in Britain and the United States, the detection of German submarines was becoming a desperately urgent matter. When a young Russian émigré, Constantin Chilowsky, suggested the possibility of an ultrasonic detection ("echo ranging") system, the French government was intrigued (Hackmann 1984, 77–81). Langevin was asked to investigate the idea and to see if a practicable device to be installed in ships could be developed. In April 1917 he succeeded in developing a quartz transmitter, using a single large and perfect crystal. His first model resonated at a frequency of 150 kilohertz, and because of the very high voltage used, it produced an enormous (1000 watt) power output. The results seem to have been spectacular: "[Langevin] reported that in trials in the laboratory tank 'fish placed in the beam in the neighborhood of the source were killed instantly, and certain observers experienced a painful sensation in plunging the hand in this region'" (Hackmann, 81).

It seems that in these early experiments Langevin was brought unexpectedly face-to-face with the fact that ultrasonic waves also had certain biological properties.[1] Subsequently Langevin changed the design of his quartz transmitter and reduced the operating frequency to 40 kilohertz. In consultation with Langevin and his team, comparable work was initiated in both Britain and the United States.

In the 1920s research on underwater acoustics continued, though on a smaller scale (Hackmann 1984, 1986). Operating frequencies were reduced further, "transducers" that could operate both as transmitters and receivers in rapid alternation were developed, and extensive studies of the transmission of ultrasound in sea water were carried out. Moreover, a piezoelectric echo-sounding device, based on Langevin-Chilowski patents, was marketed by the Marconi Sounding Device Company (Hackmann 1984, 227). In the interwar period a variety of interests in ultrasonics emerged. These included biological interests in their effects on the one hand and possible new applications of the technology on the other. One of these applications was in the realm of industrial testing, another in that of "fringe medicine."

In 1928 the Soviet physicist S. Y. Sokolov first showed that discontinuities in metal structures could be detected with ultrasound. An American scientist, F. A. Firestone, worked further on this idea during World War II, and his device went into production as a "Reflectoscope" at the end of the war. The principle, like that of "echo ranging sonar," is simple enough. Using a piezoelectric crystal as a generator of ultrasonic waves (the "transducer"), a pulse is directed into the structure. Recall that ultrasonic energy is reflected back (as is sound energy; this of course is the echo effect) when it meets a boundary between two substances. Thus if an echo is received back at the piezoelectric crystal *before* that from the back wall of the structure, then there must be some sort of internal discontinuity. Of course, there are many problems of design to a device such as the Reflectoscope—for example, one can use the same or a different piezoelectric crystal as receiver of the echo waves; and transducers having very varied operating characteristics (in terms of frequency, energy output, beam shape) are available.

In the 1930s also a practice of "ultrasonotherapy" emerged, later to become fashionable especially in Germany (on a "completely unscientific basis as a cure-all for anything from cancer to violinspielerkrampf" (Hill 1973)). This was quite separate from interest in the biological properties of ultrasonic vibrations; biological and therapeutic interests remained apart. Early suggestions for ultrasonotherapy "did not come from investigators who had been devoting their time and energy to a fundamental study of the biologic effects of ultrasound. It was here in the United States that the first research on the biologic effects of ultrasound was done. The first paper published on this subject was that by Wood and Loomis in 1927. . . . Evidently, the results of these extensive and thorough studies failed to demonstrate any promising and practical applications in medicine. . . . [In view of this] it is difficult to understand the wave of enthusiasm for ultrasonotherapy which swept several European countries during the years 1945 to 1949" (Herrick and Krusen 1954).

As the Second World War approached, interest in ultrasonic technology grew. On an increasing scale British and American warships and submarines were being fitted with "echo ranging" devices. Manufacturing output had to be increased. In the United

States, major manufacturers were the Submarine Signal Company (subsequently taken over by Raytheon) and the Radio Corporation of America.[2] In Britain they included Henry Hughes and Son and Kelvin Bottomley and Baird (later to merge as Kelvin Hughes). Important technical development in ultrasound technology occurred in the course of World War II, stimulated by the growing sophistication of submarine warfare. There was a significant research and development effort in the United States (where collaborative projects involved leading universities like Harvard, Caltech, and MIT and firms such as Bell, General Electric, and RCA) as well as in Britain and Germany. Expertise both scientific and technical in the field of ultrasound received a tremendous boost from the effort put into "sonar" and (by virtue of a certain overlap in technology) into "radar."

Perhaps unsurprisingly, given the new international tensions that were already starting to build up, military interest in sonar by no means ended in 1945. This had a number of consequences for developments in the medical area that, as we shall see, were then crystallizing. It explains Allied interest in the progress made by the Germans during the war and the steps taken to assimilate German expertise in ultrasonics.[3] It also explains efforts both in Britain and the United States to make accumulated knowledge public and to extend the range and scope both of ultrasonic R&D and of industrial involvement. In both countries contracts for the development of more sophisticated sonar equipment were given to industrial corporations in the late 1940s and early 1950s, including in the United States Raytheon's Submarine Signal Division and RCA, and in Britain EMI, GEC, Kelvin Hughes, and Plessey. Some of these firms would later show interest in exploring possible new applications of the expertise they were acquiring.

Ultrasonic Imaging: the Exploratory Phase, 1937–1953

A variety of interests in ultrasound technology built up through the 1930s and 1940s. Expertise, too, was rather widely dispersed, all the more so as military operations involved the mobilization of industry and (especially in the United States) the universities. In addition to the concerns relating to submarine detection, flaw detection, biological effects of high intensity ultrasound and "ultrasonotherapy," a new interest was slowly emerging. In the late 1930s and early 1940s a number of scientific investigators began to ask themselves a new question: Can representations of biological structures be obtained using ultrasonic beams? It is tempting to see this as connected analogically to the powerful spatial representations radar provided during the war. And although this kind of imagery undoubtedly played a part (as did the availability of technical expertise), it fails to account for the diversity of investigations that began in these early years. Investigators were led to consider ultrasonic image-formation by a variety of motivations and under remarkably different circumstances. Though major advance in the realization of a "vision" of ultrasonic imaging for medical purposes was to be slow, and would depend on expertise developed in World War II, the roots do in fact lie in the 1930s.

In the case of at least one early investigator the motivation seems to have been a general interest in the interaction of sound and tissue. The work of André Denier, redolent of an earlier era of science, was inspired specifically by his having read an account of the damaging effects of sound waves.[4] In 1940, as France fell to the Nazis, Denier had retired to La-Tour-du-Pin, a small town near Lyon in Vichy France. Having access only to the simplest equipment, he began with very simple experiments. The technical problem in devising a diagnostic technique, to which he gave the name Ultra-sonoscopie, was to find a way of detecting and transforming the outcoming sound wave in such a way as not to be affected by the nearby high frequency generator stimulating the quartz crystal. Perhaps the best approach would be to have twin quartz crystals, connected to the two plates of a cathode ray tube, combined with an impulse generator to drive the crystal in bursts. Denier saw this, but it all seemed too remote and impracticable.[5] Instead he designed a less complex method using only a microammeter to detect the current. With this basic apparatus Denier moved from simple phantoms to see what he could detect of the internal structure of the skull. Having introduced foreign bodies into cattle brains, he sought to locate them with his apparatus. Here however he found himself in difficulties. When, after the Liberation, Denier presented his work to scientific meetings in Paris (Denier 1946b), it gave rise to a harshly critical rejection from Dognon, an authority on ultrasound who had been working on its biological properties for 15 years. Though they tried Denier's methods, Dognon and Gougerot argued his experiments could not possibly be made to work (Dognon and Gougerot 1947). The weight of authority was brought to bear against the meaningfulness of what Denier had done. And that, alas, was that.

Meanwhile a more sophisticated attempt at using ultrasound to image the skull was being made in Austria. Karl Dussik, a neurologist, had become interested in the possibilities of imaging the brain and thereby locating brain tumors. In 1937, working together with his physicist brother Friedrich, he succeeded in generating an image by allowing an ultrasound beam to pass through the skull at a series of positions. The transmitted energy could be recorded as light on a photographic plate. As the transmitter was moved across the skull, according to exactly what lay in its path, the ultrasound beam would be attenuated so that patches of light and dark would be obtained. The technique was given the name Hyperphonographie (Yoxen 1987). The Dussiks had great hopes for their technique (Dussik, Dussik, and Wyt 1947). In 1948 their institute, established by then in Bad Ischl, was visited by a group of officers from the headquarters of the American army in Europe. Their report found its way to the Acoustics Laboratory of the Massachusetts Institute of Technology, which did a lot of work for the U.S. Navy and where a comparable project was just beginning.

It was at this time, with steps having been taken to make a good deal of new equipment and expertise available for civilian applications and development, that the exploratory phase of diagnostic ultrasound really got underway. In the first few years, the late 1940s, the work being done included a number of industrial projects. One of these, being carried out on a small scale at the Siemens Laboratories in Erlangen, had picked up the Dussiks' problem and was looking at the transmission of ultrasound through the skull.

In Buenos Aires McLoughlin and Guastavino were working in the laboratories of RCA's Argentine subsidiary. Having become fundamentally committed to the importance of research, RCA had become one of the major U.S. contractors for military R&D. Acoustics was one of the areas in which the firm had become preeminent during the War. In the postwar period RCA established a policy of investigating all possible ways in which its expertise in electronics could be commercialized (Graham 1986, 68–72). McLoughlin and Guastavino had become involved in assessing the possibilities of using ultrasound to detect nonmetallic foreign bodies in human organs. In 1949, in an Argentine medical journal, they described the apparatus they had constructed for this purpose, and to which they gave the name LUPAM (Localizador Ultrasonoscopico Para Aplicaciones Medicas). Having considered the various options, they had elected to build an apparatus involving a single crystal as both generator and receiver, switching rapidly between the two modes, and with the two modes connected to the plates of a cathode ray oscilloscope.[6] The system would function as a kind of radar system (McLoughlin and Guastavino 1949). With this apparatus echoes could be recorded, and an elegant series of experiments was described. Among objects successfully scanned were a human forearm and an excised kidney in which a stone had been introduced. The speculations these authors offered as to further development of the LUPAM technique are intriguing. By analogy with radar, they argued, it ought to be possible to develop a technique that would produce not merely traces but something akin to a true image. A sketch was made of what such a LUPAM image of the forearm might look like (figure 3.2). This, they suggested, might be a particularly important application of ultrasound to diagnosis. So far as I can determine, these Argentine scientists, working in an industrial research laboratory, had invented what was to become diagnostic ultrasound. Many of the technical problems seem to have been solved. But there is no evidence that the next steps were ever taken.

At the same time an American surgeon, Dr. George Ludwig, had also become interested in using ultrasound for the detection of gallstones and had done some work on this at the Medical Research Institute of the U.S. Navy. When the MIT Acoustics Laboratory people set up their medical ultrasonics project, they put together a powerful team. Working with a brain surgeon from the Massachusetts General Hospital and Ludwig (who joined the hospital in 1949), they also recruited Theodore Heuter, who had previously worked at Siemens. Having visited Dussik's clinic in October 1949, Bolt of MIT and Ballantine of the MGH returned encouraged. Funds were obtained and work began. There were clear medical justifications for pursuing ultrasonic investigations of the brain. In 1950, to obtain evidence of the distortion of the ventricular system (which may result from traumatic injury or from a tumor) was not easy. The principal radiological technique used, ventriculography, involved replacing the brain's ventricular fluid with air. Developed at the end of the First World War, it is an invasive method not without risk to the patient. Alternative techniques available, such as angiography, were little better in that respect.[7] If the necessary evidence could be obtained without employing this unpleasant and rather risky technique, it would be a major advance for neurology.

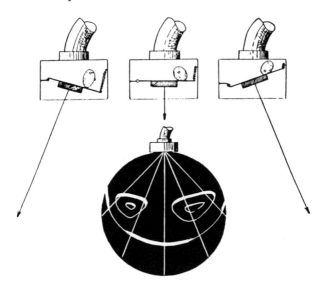

Figure 3.2
LUPAM, one of the earliest conceptualizations of the principle of two-dimensional ultrasonic imaging. The diagram is an imaginary representation of the human forearm. From McLoughlin and Guastavino 1949.

The Boston group set out to replicate Dussik's technique and to embed it in an ambitious program of broader scope.

By spring 1950 they had succeeded in constructing a scanner like the Dussiks' (Yoxen 1987), and at the end of 1950 the MIT group explained its rationale and basis in an article in *Science*. They saw themselves as engaged in a "long term program on the application of ultrasonic techniques to medical problems" with as "immediate goal the detection and localization of intracranial tumors and cerebral anatomic abnormalities" (Ballantine et al. 1950).

The MIT/MGH group had elected to continue with Dussik's transmission method—detecting the ultrasound passing through the skull—although they were well aware that the reflection method, which lay at the heart of naval sonar as well as industrial flaw detection equipment, offered another possible approach. They had shown, at least to their own satisfaction, that the transmission approach was the more promising for brain imaging. Others, however, were beginning to look into the diagnostic possibilities of the reflection technique. In *Science* and later in the *Journal of the Acoustical Society of America* the MIT Acoustics Lab people argued that their transmission method, based on fundamental studies of interaction between brain tissue and ultrasound, should work better than the other approach.

By the end of this exploratory period in the history of diagnostic ultrasound they had been forced to cede the ground to their erstwhile rivals. Despite the quality of the supporting biophysical work being done, diagnostic results were disappointing. Theoretical prediction to the contrary, the effects of the skull itself on the sound beam were proving much greater than those of the ventricular structure being investigated. As a result images were poor. In Erlangen, Güttner and his colleagues had concluded in 1951 that transmission imaging of the brain by the Dussik method was impossible (figure 3.3) (Güttner et al. 1952). A conclusive test was required. This was achieved by incorporating a diagnostic x-ray unit into the ultrasonic head scanner so that images of precisely the same area would be obtained by each method. When this was done with patients suspected of having brain tumors and the two sets of images were compared with the results of postoperative evidence, it became clear that the x-ray method was the more accurate. By 1954 it began to seem doubtful "in view of the anatomy of the human head, whether transcranial ultrasonograms of the Dussik type . . . can yield reliable information on intracranial structures" (Ballantine, Heuter and Bolt 1954). In this paper the Boston scientists announced that they were terminating their line of investigation. Though improvement in instrumentation might improve the quality of the results, because of the parallel development of other techniques (not only the ultrasonic reflection method was mentioned but also radioisotope scanning, which was then being developed), they decided to abandon their attempts at imaging. The future rested with those investigating the reflection method.[8] There were two important research efforts to start with, each bearing the mark of the medical specialty within which it was conceived.

In Denver, Colorado, the radiologist Douglass Howry had begun work in 1948, shortly after graduating from the University of Colorado School of Medicine (Holmes 1974, 1980). On the basis of some preliminary studies Howry rapidly reached two important conclusions. First, he discarded the transmission approach in favor of the echo or reflection method. Second, he was not impressed by the results of trying out an industrial flaw detector directed into the body. To be sure, echoes could be obtained with the Reflectoscope, and when projected onto a cathode ray oscilloscope they "appear

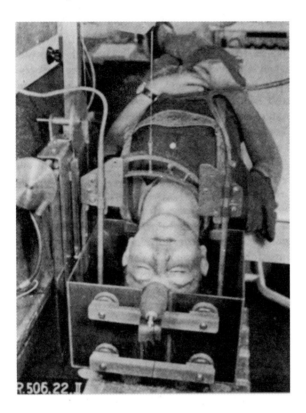

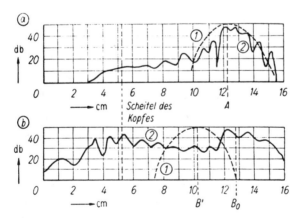

Figure 3.3
Experimental apparatus designed to test the feasibility of transmission imaging of the brain, built in the Siemens Laboratory in Erlangen. From Güttner 1952. Photograph courtesy of S. Hirzel Verlag and the Deutsche Physikalische Gesellschaft.

much like the waves of an electroencephalogram." However application of this ultrasonic phenomenon to diagnostic medicine did not seem very promising "because the complexity of the sound-reflecting structures produces echoes which are so numerous and variable that controlled experiments and consistent results are difficult to obtain" (Howry and Bliss 1952). As a radiologist, Howry was very well aware of the limitations of conventional x-ray technology when it came to imaging organs such as the liver, the pancreas, and the female breast. The densities of adjacent soft tissues (and hence their effect on x-rays) were often so similar that boundaries could scarcely be seen, consequently the technique was of limited use. What Howry was looking for was not a technique for neurology (as Dussik or Ballantine were seeking), but a means of extending the scope of radiological practice to include the effective imaging of soft tissue. According to Howry, "The value of such a system would be greatly enhanced if a true image similar to a roentgenograph or photograph could be obtained, for it would then be possible to visualize normal and pathologic soft tissue structures, nonmetallic foreign bodies, nonopaque gallstones, etc." (Howry and Bliss 1952).

Working together with W. R. Bliss, an electrical engineer with the University of Denver's Institute of Industrial Research, Howry set about designing a different kind of apparatus. In their 1952 paper, published in the *Journal of Laboratory and Clinical Medicine*, Howry and Bliss explained the principle of their "somascope" (somascope implying tissue-vision). We learn here that Howry had chosen to work at a frequency of 2 Mc/s and that he had built in a means of focusing the sound beam. Figure 3.4 from this paper shows the apparatus and its principle schematically. The transducer was placed in water to reduce loss of energy; the test object was also in water. Howry and Bliss used a visual image to show their readers how the instrument worked. The test objects, a condom filled with fluid and another filled with a water/talc mixture into which a glass rod had been placed, provide familiar enough images to make interpretation clear for the reader. As its pièce de resistance the paper contained a "somagram" of the forearm of one of the authors, together with a corresponding anatomical drawing (see figure 3.5). Here "both the ulna and the radius are very clear," though the two bones cast a sonic shadow and "the loss of all signals behind them is apparent." Howry was confident enough to conclude that the

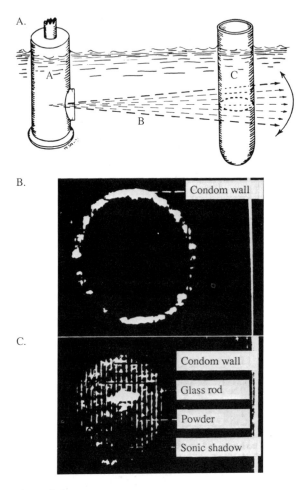

Figure 3.4
Experiments carried out by Howry to establish the feasibility of ultrasonic echo imaging. From Howry and Bliss 1952, 583.

ultrasound technique would eventually prove its worth to the radiologist as an important "adjunct to x-ray in diagnosis."

The second program of research into reflection ultrasound had begun in Minneapolis, Minnesota. An English surgeon, John Wild, working at the University of Minnesota, began to collaborate with Donald Neal, an aeronautical engineer working at a nearby naval air station. Wild also came into contact with F. J. Larsen, the author of *Ultrasonic Trainer Circuits* and affiliated with the research department of the Minneapolis-Honeywell Regulator Co. It was an apparatus Larsen had helped design, constructed to train pilots in the

TRANSVERSE SECTION THROUGH MID-FOREARM

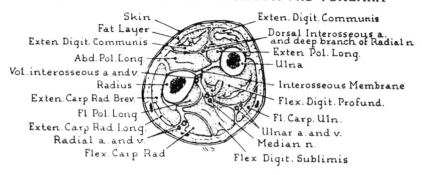

Skin
Fat Layer
Exten. Digit. Communis
Abd. Pol. Long
Vol. interosseous a. and v.
Radius
Exten. Carp Rad. Brev.
Fl. Pol. Long
Exten. Carp Rad. Long
Radial a. and v.
Flex. Carp Rad

Exten. Digit. Communis
Dorsal Interosseous a. and deep branch of Radial n.
Exten. Pol. Long.
Ulna
Interosseous Membrane
Flex. Digit. Profund.
Fl. Carp. Uln.
Ulnar a. and v.
Median n.
Flex Digit. Sublimis

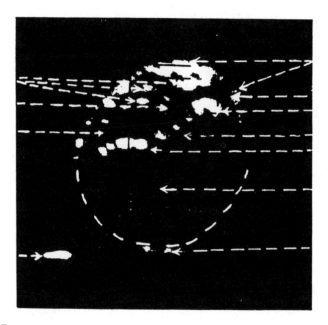

Figure 3.5
Howry's first published somagram of the human forearm, with anatomical drawing for comparison. From Howry and Bliss 1952, 589.

use of radar, with which Wild began his work.[9] As a surgeon Wild shared neither the neurological priorities of Dussik and Ballantine nor the radiological perspective of Howry. He saw in ultrasound a possible means of distinguishing normal from cancerous tissue, and from in vitro studies of tissue he moved rapidly to look at cancers. In the first instance this problem of tissue differentiation did not appear to necessitate images like those of roentgenography, and Wild's first apparatus simply produced the wavelike pattern that Howry had found confusing (Wild and Neal 1951). The transducer (crystal) is electronically caused to resonate at a frequency of 15 megacycles per second, vibrating for 1/2 of one millionth of a second and then stopping, producing pulses of ultrasound. The ultrasound waves pass through water contained in the chamber where the crystal is located and thence through a rubber membrane closing off the chamber. The biological tissue sample is in contact with this membrane. Between the tiny vibrations the crystal stops, functioning in those periods as receiver in place of generator. The frequency at which the ultrasonic waves are generated (in this case 15 megacycles, or Mc/s) was known to be an important parameter. The higher the frequency used, the greater the resolution (or separating power) obtained, but the smaller the penetrating power. In choosing to work at 15 Mc/s, (and thus aiming for high resolution) Wild and his coworkers were setting a crucial constraint on their work, effectively limiting themselves to superficial sites in the body.

Wild began with a number of experiments before he felt able to turn his apparatus onto living patients. Could the apparatus distinguish between normal and cancerous tissue? Pieces of brain tissue were placed on the membrane of the apparatus: first normal tissue, and then tumor tissue (a thinner slice, since the rays did not penetrate it so well). It seemed to work. Would it work in whole brains? The brains of a number of deceased patients, believed to have died from brain tumors, were removed at death and examined. Here too results were promising (figure 3.6). It did seem possible to detect various kinds of neoplasms (tumors) in the brains. The third question related to the safety of ultrasound. It was known that high-intensity continuous ultrasound could produce tissue damage. What of the low-intensity pulsed ultrasonic beam they were using? Experiments were carried out on animals: four rabbits and a cat. Their brains were exposed to ultrasonic vibrations from the instrument for periods of 20 minutes and then examined

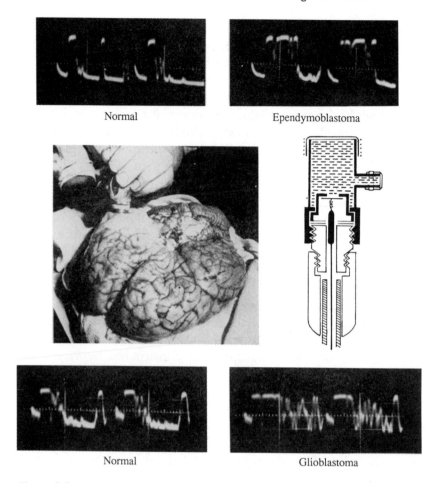

Normal   Ependymoblastoma

Normal   Glioblastoma

Figure 3.6
Wild's echoscope, showing its application to the study of brain tissue. From
Wild and Reid 1953. Photograph courtesy of the American Institute of Physics.

for evidence of neurological damage. None was found. Subsequently, the animals were "sacrificed at periods of 4 hours, 2 weeks,
6 weeks, and 8 weeks after exposure," and histological tests were
carried out. There was no evidence of tissue damage. These results
were widely publicized (for example, in 1951 issues of both the
Journal of Neurosurgery and the *Journal of the American Cancer Society*).
There seemed reason to believe that the technique might work and
that it was harmless. On the basis of this series of experiments, Wild
moved on to his first clinical studies.

Although he had been working principally on the detection of cerebral tumors, for this first clinical study Wild turned to tumors of the female breast. Two women, each with a lump in the breast, were successfully examined. Comparison between the affected and the normal breasts suggested a carcinoma in the one case, a simple cyst (fibroadenoma) in the other. Biopsy confirmed these diagnoses and, by the same token, the value of the technique. But why had Wild moved from the brain to the breast? This was implicitly justified (in a later paper) by slightly redefining the object of the work: in place of detection of "cancer," the focus is now adjusted to the limitations of the apparatus. "The primary objective of the cancer studies" now becomes "the diagnosis and possible detection of cancer in common accessible sites all over the body, such as the upper digestive tract, the lower bowel accessible from the anus" (Wild and Reid 1953). "Accessible" cancers became the focus of attention, and the tumerous female breast was chosen as "representative" of these. The female breast was to remain the principle focus of attention in Wild's ultrasound work.

In 1952, with support from the National Institutes of Health, Wild moved his work out of the naval air station and into the hospital. A new instrument, an echograph, was constructed by Wild's co-worker, the electronic engineer J. M. Reid. Continuing with the 15 Mc/s operating frequency, the object was to get "a working machine for clinical use as quickly as possible." Further breast cases were also successfully examined, and what was needed was an "objective" diagnostic criterion. Wild needed not only to be able to assess the value of his instrument, but if all went well he also had to convince professional opinion of the reliability of its diagnoses. Statistical examination of the pairs of echograms for 21 patients suggested "ratios of the areas under traces" for a given patient as a diagnostic criterion (see Wild and Reid 1953).

Before the significance of this quantitative criterion had really been established, a major change in *technology* led to a reevaluation of what had thus far been achieved. In fact, Reid had begun work on a new ultrasonic device some time before. In place of the one-dimensional traces, the objective in the new instrument was a two-dimensional record, more akin to a picture. A two-dimensional map on which the tumor could be located would greatly facilitate diagnosis, and this was reported in *Science* in 1952 (Wild and Reid 1952b).

The apparatus had been modified so that the crystal, still contained in a water-filled tube closed off by a rubber membrane, could be rotated. The apparatus was designed so that its rotations were synchronized with the cathode ray tube. In a demonstration experiment a 1 cm cube of beef kidney was placed on the rubber membrane. By displaying the scans of this test object, the two essential modifications could be simply explained. The old one-dimensional traces could be displayed as a set of light spots, of intensity corresponding to the strengths of the echoes. The output is a line of spots of variable brightness. If the crystal is slightly rotated, the beam of ultrasound traces a different path through the test object and a second line of spots is obtained. In this way a set of these variable intensity lines of light spots could be traced out on the cathode ray tube through rotating the crystal. This set of lines was called a "sector scan."

By 1953 both Howry and Wild had obtained ultrasonic images. However—and this is crucial to understanding the future course the "career" of ultrasound would take—they had derived their images from quite different research programs, and they attached different significances to them. For Wild, the surgeon, the crucial point was to translate the area-ratio criterion with which he could recognize cancer into the new language of the images as quickly as possible. (By late 1953 he had succeeded. A malignant tumor, returning more sound, would appear as an area of brightness. To show the relevance of the the technique for cancer diagnosis, a series of four breast cancer cases was submitted to the journal *Cancer Research*.) Howry was at a different stage in his attempt at developing an imaging modality of general application within radiology. In pursuit of this objective he was asking "What sorts of objects can I image with this equipment" and "What are the problems with my imaging technique?" It is from his radiological perspective that Howry's need to understand and eliminate factors that limited the quality of the images, such as the "sonic shadow," derives. In 1953, when a new "somascope" was completed, it was said that "The ultimate use of this equipment would be in the living patient. Because of the temporary nature of the construction of this early experimental model, it is possible to examine only the extremities of living subjects. There is no sensation associated with its use and, in more than four and one half years of constant association with this equipment, no injurious effects have been noted by any of the group" (Howry, Stott, and Bliss 1954).

The exploratory phase of diagnostic ultrasound had begun with the neurologist Dussik at the end of the 1930s. Theoretical prediction to the contrary, by 1953 Dussik's transmission approach (detecting the pattern of rays passing through the body) had been discarded in favor of an echo method (in which the pattern of reflected rays forms the trace or image). To that degree the future course was set. Nevertheless, the emerging technique was already taking two quite different forms. Howry's work and Wild's work were giving rise to two designs being shaped and reshaped within two quite distinct sets of medical interests. We can see various signs of this. One is the way in which, subsequently, the two men would look to demonstrate in print the significance of what they had achieved. By 1956 Wild had enough data to begin to work out some statistics expressing the accuracy with which breast lesions could be diagnosed. The task for him was clear—cancer detection—and what had to be demonstrated was that the technique could do the job. Howry on the other hand was after an imaging technology that would be put to the very varied uses constituting general radiological practice. He based the claims made for the somascope not in diagnostic statistics but by emphasizing to the quality of the image his instrument produced. A second difference is to be found in the diagnostic criteria implicitly or explicitly informing the work. The difference comes strikingly out of the following exchange that took place at a 1955 conference (Kelly 1957, 63–64):

Dr. Howry: "One thing I neglected to say. I have not in any way contradicted Dr. Wild and Dr. Reid's work, because we have not investigated the same problem as they have. They are most assuredly operating over a full 60-decibel range. They are not trying, apparently, to get the last ounce out of definition, which is what we are trying to do."

Dr. Wild: "I would like to add that we are attacking the known sites of cancer and holding the cancer up, so to speak, on the end of a rubber membrane."

Dr. Howry: "You are comparing the picture of one type against another which is being called standard, and from that arriving at the information—without making any definite statement that it is the true picture of the tumor. I think that from that standpoint it is quite clear. You are certainly getting significant results."

In these different understandings of diagnosis and the different medical communities with which they are associated lie the roots of the complexity characterizing ultrasonic imaging's incorporation into the diagnostic armamentarium of modern medicine.

Ultrasound's Development Phase, 1954–1965

The next stage in the career of ultrasound was not to be a contest between the visions of these two men. Wild's work was soon to be discontinued. A period of growing concern regarding breast cancer and widespread demands for screening programs provided an opportunity for a major extension of their practice that was seized by radiologists. The specialized x-ray technique known as mammography was developed for breast examination in this period. Moreover Wild's experimental choice for 15 Mc/s was to prove highly problematic and would come to be seen as mistaken by other workers. Yoxen refers to Wild's research funding being terminated and subsequent "massive litigation between him and the agency managing his research" (Yoxen 1987).[10] On the other hand, as ultrasound very gradually approached the stage of commercialization, it did not become quite the addition to the radiologist's armamentarium that Howry envisaged. This next phase in the career of ultrasound is marked by the loosely coupled constitution of the technique within a still greater variety of clinical contexts, each with its own requirements, clinical agendas, and specialist audiences. Cardiology, neurology, ophthalmology, and obstetrics and gynecology were to prove scarcely less significant than radiology as medical contexts for ultrasonic diagnosis.

The development phase in the career of a technology demarcates the process by which a technology becomes a commercial product, so that the involvement of manufacturing industry is a central feature. How does industry become involved, what industry, and on what basis—these are the sorts of questions that must be posed. In the case of diagnostic ultrasound the development phase proves to be characterized by the *variety* of medical-industrial partnerships that evolved. Typically these partnerships grew out of apparently chance encounters within particular local milieux. As we follow the career of diagnostic ultrasound further, we shall see that industrial participation corresponds to, and thus may have consolidated, the variety that we find in medical contexts. Thus, although some x-ray suppliers did become involved, firms that looked to redeploy specific expertise in acoustic technology—often derived from experience with military or industrial devices—were, generally speaking, well able to compete.

In the mid fifties much of the time of Howry and his colleagues was taken up with design features that limited the capacity of their somascope to produce images of the quality they sought. One was the shadow effect. Another was the false echoes obtained from "structures of fairly simple geometry [which] present surfaces with steep angular sides." "False echoes" were being obtained "derived from the sound ricocheting around inside the body much like a ball on a billiards table." This problem was investigated experimentally with various test objects. A further problem was an effect whereby the echo information received depended upon the angle at which the sound beam struck the object. Small departures from perpendicular have serious effects on the sound returned and so on the signal strength, a matter with important implications for the possibilities of imaging curved structures.[11] Although by 1955 a remarkably detailed image of a curved structure—the neck of one of their engineers—had been obtained, this had entailed his sitting immersed to the neck in a tank of water—and it had been possible only by making four separate pictures from the 90 degree quadrants of the neck. Howry's research agenda was set by his concern to improve the imaging quality of the apparatus, in line with the conventions of radiology. He reported, "Our recent work, therefore, has primarily been the development of an instrument which allows the sound head to scan horizontally back and forth, and simultaneously travel round the structure under study which corrects for the above deficiencies. This type of complex picture registration we choose to call a "compound circular scanning system" (Howry 1957).

By 1957 Howry and his colleagues had completed a third-generation somascope, which they had built from a B-29 gun turret (figure 3.7). The transducer, immersed in water, traveled around a metal ring, running full circle around the gun turret/scanning tank. Simultaneously, the transducer moved back and forth in a four-inch linear scan. This "compound circular scanning system" proved to be an effective means both of eliminating artefacts and of seeing around curved structures. By 1956, using this new apparatus, it was possible to make as good a scan of the human neck as had previously been made by pasting four quarter scans together (Holmes 1980). The technique being used still made use of a water bath to ensure good acoustic coupling between the transducer and the patient's skin, but the patient no longer had to sit immersed in water. The new design used a semicircular pan filled with salt water against which the patient, seated in a dentist's chair, could be moved in the vertical plane. In this way serial cross sections at different levels

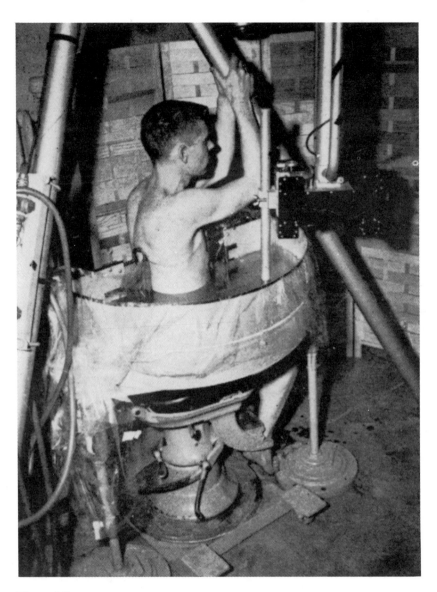

Figure 3.7
Third-generation Somascope showing a patient seated on a dental chair, which can be raised and lowered in the semicircular, fluid-filled pan. From Holmes and Howry 1963, 15. Photograph courtesy of Plenum Press, N. Y.

through the body could be obtained. "Time-varied gain" was employed, so that the sensitivity of the receiver increased from the skin surface.[12]

By the early 1960s Brown and Donald in Glasgow built a compound scanning device that achieved acoustic coupling without the use of the cumbersome water path. In 1960 work on a similar device began in Denver, and by 1962 William Wright had constructed a contact scanner in the University of Colorado Medical Center. This was described in print in 1965 (Holmes et al. 1965). Designed to operate variably at between 1 and 2.5 Mc/s, the instrument was conceived specifically for examination in the abdominal region.

Gradually the range of clinical practices within which the diagnostic possibilities of ultrasound were being investigated was growing. Access to the technology was deliberately stimulated immediately after the war, and the idea that it might in some way or other be useful in diagnostic medicine was apparently in the air. There was little attempt deliberately to replicate the work of others. Many of the fifteen to twenty investigators who began work in the 1950s had no knowledge of Howry's work, and the majority of them were not radiologists. Looking across the range of medical contexts, we find that in some cases it proved possible to make considerable progress in diagnostics using the industrial flaw detector, scarcely modified. Under these circumstances little effort was put into technical development work. In other cases the complexities of the echoes obtained rendered the flaw detector of very limited use. Two examples exemplify these processes and at the same time show the local contingencies that mark this stage of ultrasound's constitution.

The first story is set in the small university town of Lund, in southern Sweden. It begins in 1950, when Lars Leksell, Professor of Neurosurgery, borrowed a Kelvin Hughes industrial flaw detector. Leksell had for some time been on the look out for a means of rapidly and noninvasively locating a hemorrhage in the brain caused by a blow to the head. This would greatly facilitate the task of the neurosurgeon faced with the need to open the skull for operation. Leksell had the idea of using the ultrasonic flaw detector for this purpose. However, his experiences were disappointing and the instrument was returned. Nothing further happened for three years.

In 1953 a series of remarkable coincidences occurred. Dr. Inge Edler, responsible for preoperative diagnosis in cardiology, met by chance a graduate student in the university's department of physics, Hellmuth Hertz. Hertz was working in nuclear physics but was also interested in ultrasonics. He was therefore intrigued by a suggestion made by Edler in the course of this meeting. Edler wondered aloud whether something like radar might be of any help with a particular problem of cardiological diagnosis with which he was preoccupied. Hertz later described what happened next:

Luckily, at that time I knew that the first ultrasonic reflectoscope for flaw detection had been delivered to the Tekniska Roentgencentralen, a company in Malmo which was responsible for the nondestructive testing at the large shipyards in that city. After consulting my medical student friends about the position of the heart, I went to Malmo to see if any echoes could be obtained from my heart with that equipment. Because of the positive result of that experiment Dr. Edler and I contacted the company who kindly agreed to let us have the reflectoscope for a weekend at the hospital in Lund. An intensive study during those two days convinced us that the ultrasonic reflectoscope method could become a valuable tool for heart diagnosis and we decided to carry out further studies. (Hertz 1973)

Edler and Hertz lacked the resources to purchase a reflectoscope for their further studies. That they were able to obtain one was due to the curious fact that Hertz's father had been director of Siemens research laboratories before the end of World War II. Hertz was able to contact the vice president of Siemens Medical Division in Erlangen and through him secure the loan of a Siemens flaw detector. This arrived in October 1953, and studies focusing on heart structure/ motion as a function of time were started.

Lund was a small place and it was not long before Leksell learned of the presence of the Siemens instrument. In December 1953, faced with an emergency operation on a child in which it was unclear on which side to operate, Leksell decided to try ultrasonics once again. He borrowed the instrument. It worked. Leksell was able to see to which side the brain had been shifted by the suspected subdural hematoma and to plan his surgery accordingly. His interest in ultrasound was now rekindled and he continued his investigations with a later Kelvin Hughes instrument. In the course of 1954 Leksell announced his success at a Swedish conference, and some of his results appeared in print two years later (Leksell 1956). By this time the Lund workers knew of the other work in the area, and Leksell was aware that others (Güttner, Heuter et al.) had come to doubt the possibilities of ultrasound in brain diagnosis.[13] Perhaps it ought not to work, but it did. According to Leksell, "The exact anatomic basis for the different echoes from the brain has not been elicited as yet. The cases presented in this paper show, however, that the echo method can give valuable diagnostic information with the scalp and skull intact" (Leksell 1956).

Leksell had been led to try the method once again by a clear clinical need—"a rapid and safe method for exploring the intracranial contents in acute head injury"—where available alternatives were either inaccurate or risky or both. His approach to ultrasound investigation of the brain was very different from that of Dussik or the MGH group, and Leksell was aware that he could say nothing of the applicability of the method to nontraumatic disorders, such as tumors. In any event a clear clinical use had been rapidly established in neurosurgery, and the commercially available flaw detector, producing a measurable one-dimensional trace, had begun to earn its place in the clinic.

In cardiology the situation was somewhat different, and Edler's interests were initially less precise. However he rather rapidly concluded that the principal use of the technique would probably lie in diagnosis of incompetency of the mitral valve of the heart (mitral stenosis). By placing the transducer over the heart, vertical deflections due to the heart's movements could be obtained. By analogy with electrocardiograms, or ECGs, (with which they proved to be synchronized) the new traces became known as ultrasonic cardiograms, or UCGs (Edler 1955). These UCGs could be characterized by reference to the ECGs; the forms of both relate to specific motions of the normal heart.

In both neurology and cardiology these Lund workers had established clinical uses for ultrasound, making use of the commercially available flaw detector with its "A mode" (or one-dimensional) trace. In their own clinics, both Leksell and Edler were making routine use of the instrument. This is of course not to say that the clinical use of ultrasound in these specialties had been established, other than locally. Nevertheless, through publications attention was being drawn to the possibilities of the technique. Its clinical use was beginning to look both worthwhile and interesting within both neurological and cardiological fields of medical practice.[14]

The second story is quite different. Set in Glasgow, it relates to the origins of gynecological and obstetrical uses of ultrasound. In striking contrast to the situation in neurology in particular, here it was by no means a matter of a clear clinical need; rather it was an attempt to see what, if anything, the technique could do. Before leaving London in 1954 to take up the Chair of Midwifery (gynecology and obstetrics) in the University of Glasgow, Ian Donald met John Wild "whose early work with Reid in Minneapolis I already knew about" (Donald 1980).[15] Having become intrigued by the possibilities of ultrasound, on arrival in Glasgow Donald set out to investigate its properties. He began by looking at the hemolyzing effects of high-powered ultrasound on blood—a far cry from diagnosis. As for how he eventually came to look at diagnostic possibilities, Donald has provided a graphic account:

By good fortune, the wife of an eminent engineer at this time gratefully survived a hysterectomy at my hands and introduced me to her husband. Following a lunch party with their research directors a session was arranged in the research department of Messrs Babcock & Wilcox. I shall always remember that hot sunny afternoon of 21 July 1955 when we took to the factory some selections of the last few days operating, in the boots of two cars, large ovarian cysts and uterine fibroids, calcified and plain. The firm very thoughtfully provided a truly massive piece of prime steak as a control material.

All I wanted to know, and this was surely not asking too much, was whether a metal flaw detector could show me on A-scan, which was then all we had, the difference between a cyst and a myoma and so forth. To my surprise and delight the differences were exactly as my reading had led me to expect.

Donald goes on to recount how Babcock & Wilcox put him in touch with the nearby firm of Kelvin Hughes, which had made the flaw detector that had been used. Kelvin Hughes was by then developing some medical contacts, notably among neurologists interested in following up Leksell's work. Donald acquired a machine that was being used without success at the Royal Marsden Hospital in London. In Glasgow too its use was not without difficulty. "As far as I remember," Donald goes on, "it had a 1 or 1.5 MHz quartz crystal and a paralysis time of 8 cms, i.e. nothing showed up within the first 8 cms of penetration. This led us, inevitably, to devising water tanks with flexible latex bottoms which were applied with a film of grease to the protuberant female abdomen and into the surface of which our probes were cautiously dipped. Accidental spillages were frequent which such precarious balance and resulting wet beds endeared me to neither patients nor nurses."

There were some successes in the gynecological area. Despite the initial skepticism of Donald's colleagues, "Soon I was being asked to do the impossible but fortunately a brilliant young man, Tom

Brown, who was on the research staff of Kelvin Hughes, came on the scene." Brown had experience with industrial ultrasound; when he saw the pattern of echoes Donald was getting, he realized that it was far too complicated for meaningful interpretation. The clinical practice of the gynecologist was imposing far greater demands on the apparatus than did those of the neurologist or the cardiologist. If ultrasound was to have a significant role here, Brown reasoned, then an innovation in design was needed. Perhaps the way forward would be to try to relate the echoes to each other spatially. As in the cases of Howry and then Wild the notion of two-dimensional ultrasonic representations (images) was emerging once again, though in relation to yet another set of clinical priorities and constraints. By 1956 Brown had built a prototype "B-scan" instrument for scanning the abdomen.

The object was, of course, to position the echo dots geometrically in accordance with the attitude of the exploring crystal on the patient's body surface and to relate them all to their points of origin within the body. This all sounds remarkably obvious today but, believe me, we hardly knew at that time how or where to look. . . . We acquired a lot of experience and were able to give a demonstration on a patient to a meeting of the American College of Surgeons, which was held in Glasgow. At this time in 1958 we could differentiate with reasonable certainty between quite a variety of gynecological tumors and ascites. . . . We could also demonstrate fetal echoes in utero.

As the instrument was gradually adapted to the needs of gynecological practice, further improvements were made. Kelvin Hughes management was persuaded to make further commitments by the strong support the gynecologists were clearly giving to Brown's work. Having obtained the necessary backing, Brown spent half his time with the physicians, gathering the knowledge needed for further refinements. With the original flaw detector adequate acoustic coupling with the abdomen had been possible only through use of a water filled tank. This limitation was overcome when it was found that air could be eliminated and adequate coupling achieved by means of a thin layer of oil applied to the skin. A "contact compound scanner" was emerging, and with it the team of Donald, MacVicar and Brown gradually moved to more complex gynecological cases and to obstetrics. The first report on this work, based on examination of 100 patients, appeared in the *Lancet* in 1958. Here is how use of the new technique was described: "The probe is moved slowly from one flank, across the abdomen, to the other flank, being rocked to and fro on its spindle the whole time to scan the deeper tissues from as many angles as possible. At present this is done by hand, but we have plans for mechanical scanning, which should produce more consistent results" (Donald, MacVicar, and Brown 1958). The claims made were modest: "The fact that recordable echoes can be obtained at all has both surprised and encouraged us, but our findings are still of more academic interest than practical importance, and we do not feel that our clinical judgement should be affected by our ultrasound findings."

It took another year, Professor MacVicar recalled, before the group was convinced that it had an instrument of general gynecological utility.[16] This conviction set the agenda for their clinical research, as a variety of gynecological and then obstetrical uses were explored. At the same time Donald was keen to share his appreciation of the possibilities of ultrasound as widely as possible, more than willing to lend apparatus wherever interest was shown. So far as obstetricians were concerned, it is important to note that this was a time of considerable sensitivity to the risks associated with obstetrical radiography,[17] so that the time was in a sense ripe for ultrasound. By the early 1960s not only qualitative but also quantitative obstetrical work was yielding satisfying results, with the first comprehensive account of fetal skull measurement by ultrasound appearing in 1964.[18] Clinical value seemed clear. Measurement was to be a major contribution of ultrasound to obstetrical practice.[19]

What of safety? Given concern at possible dangers of obstetrical radiography, obstetricians might be expected to have been sensitive to the issue, with investigation of the effects of ultrasound a major issue on their research agenda. By the beginning of the 1960s Donald in Glasgow and Howry and Holmes in Denver were engaged in significant abdominal ultrasound studies, and claims regarding visualization of the pregnant uterus were beginning to emerge. At the same time the stage of commercialization was approaching. Were the researchers themselves convinced that the technique was safe enough? On the basis of what evidence? In a contribution to the *American Journal of Digestive Diseases*, having something of the character of a review, Howry and Holmes devoted one paragraph (roughly 2 percent of the length of the article) to the safety of ultrasound (Holmes and Howry 1963). They asked "Is there any danger to the patient from ultrasound as employed in diagnostic studies?" Evidence presented was scanty: a report from Baum and Greenwood that no effects on the eye had been observed; the fact that their own research staff was seemingly unaffected; and a "preliminary toxicity study" carried out on pregnant rabbits that showed no effects on newborn offspring. Thus, "on the basis of the evidence available," they write, "there are no toxic effects from currently used ultrasonic diagnostic techniques," and the technique may be presumed safe. Donald, by contrast, felt it necessary to engage the help of an anatomist, and he defined the problem somewhat differently. Two day-old kittens, anesthetized, were exposed to one hour's continuous radiation to their skulls , with two other kittens as controls. One experimental and one control kitten were killed after 24 hours, and the second of each group was sacrificed after 3 weeks. With the help of an anatomist, the kittens' brains were removed, fixed, sectioned, and examined. All tests for damage to the brains were completely negative. Work could go on, though Donald was later to refer to the difficulty of mounting proper safety studies. In fact, though clinical research agendas were largely being set within the distinct specialist practices, investigators were quite willing to rely on one another's findings, however tentative, regarding safety. When we later note subsequent concern with the safety of ultrasound, it will be useful to bear the nature and limitations of these experiments in mind.

By the early sixties diagnostic ultrasound devices were coming onto the market, first slightly adapted A-scan instruments specifically intended for neurological, ophthalmological, and cardiological use, and later B-scan devices. For firms making flaw detection equipment or otherwise involved in ultrasonics, producing a medical A-scan device, for example for ophthalmic work, was a relatively simple matter. The barriers to entry were extremely low.[20] Manufacturers of industrial flaw detectors such as Sperry in the United States, Siemens and Krautkrämer in Germany, and Kelvin Hughes in Scotland were very willing in the first instance to lend instruments for medical studies. Kelvin Hughes can serve as an example of how the processes might typically have unfolded. Principally involved in the manufacture of marine instruments, and a major contractor in the area of naval sonar, in the late 1940s the firm had started producing ultrasonic flaw detectors. It proved a difficult market for them, and when in the late 1950s reports began to suggest that the instrument might also have uses in neurology, Kelvin Hughes was interested. A link was formed with a neurologist, Jefferson, who carried out a number of studies using their flaw detector. Little further development work was required. This instrument, only slightly modified, was put on the market, and between one and two hundred were sold.

The subsequent collaboration between Donald and Kelvin Hughes was a very different matter and illustrative of the complex processes, both technical and social, which the adaptation of ultrasound *also* entailed. Gynecology and obstetrics led to different demands being made of the apparatus and thereby to a far more complex partnership. By virtue of the protracted technical work required, the stakes were raised. Subsequent events show how, in order to understand the development of this more complex ultrasound technology, we have to attend to external circumstances.

Brown's prototype B-scan instrument, developed in collaboration with Donald and MacVicar, was completed in 1956. At this point Kelvin Hughes was taken over by a larger, London-based firm, now known as Smiths Industries. At first this had little consequence: Kelvin Hughes retained its own Glasgow-based board of directors. Moreover, Donald was able to reduce the (financial) risk the project entailed by raising outside funds to support it. The joint project with Donald looked like a great success, and from the firm's point of view it was very fortunate that they were working together with an obstetrician. "The one area where we could make it pay off. . . . Professor Donald was in that."[21]

In 1958 the first results of Donald's work were published in the *Lancet*, as we saw, leading soon afterward to Sundén's visit. On Sundén's return the University of Lund, with funds from the Swedish Medical Research Council, ordered an instrument like Donald's from Smiths Industries.[22] This was the very first order. The instrument delivered to Sweden in 1961 (figures 3.8, 3.9) included a number of modifications to the first Donald/Brown instrument, such as the possibility of rapid switching between A-scan and B-scan. But the Smiths engineers were still not totally happy with the design. There were features that were felt to make production of the device uneconomic. Further development work was carried out and (particularly important, in reducing the uncertainties associated with taking the next step) the British Ministry of Health placed an order for five machines. With this order in hand, plans were laid for manufacture. No one could have foreseen how long realization of these plans would demand.

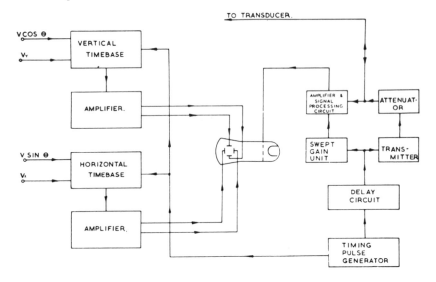

Figure 3.8
Diagram of the first commercial ultrasonic imaging device delivered by Smiths Industries to the Department of Gynecology and Obstetrics of the University of Lund in 1961. From Sunden 1964, 39.

Unlike Kelvin Hughes, Siemens was a long-established manufacturer of medical equipment, including roentgenographic. The company had a more complex and discontinuous pattern of involvement with diagnostic ultrasound. After Güttner had shown the impossibility of transmission imaging of the skull in 1952, Siemens lost interest in diagnostic ultrasound. Their reentry to the field was delayed by another circumstance.[23] About 1950 both Siemens and the small Krautkrämer company, based in Cologne, started to make flaw-detecting equipment. Located close to the steel industry—the main users—Krautkrämer could provide better service than Siemens, and it came to dominate this market. Around the end of 1956 Siemens decided to stop producing flaw-detection equipment. It was Hertz who rekindled their interest in ultrasound technology. Visiting the company in 1957, Hertz adapted three of their flaw detectors for heart investigations, and these were placed in hospitals in Germany. Hertz related, "However, only in one hospital (Medizinische Klinik in Düsseldorf) a good physician (Dr. Sven Effert) used it with appreciable success, thereby creating an early start for echocardiography. At the other places the method died

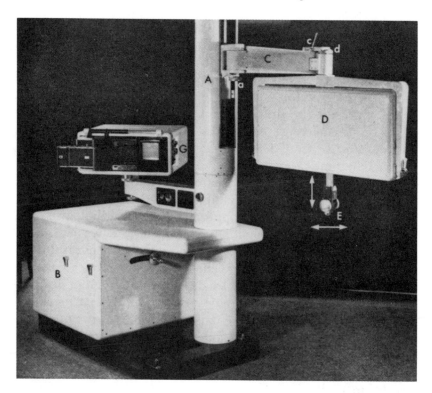

Figure 3.9
Device delivered to the Department of Gynecology and Obstetrics of the
University of Lund in 1961.

due to disinterest. During my stay at Siemens we developed some
special electronics for ultrasonic work. This work was carried on
after my departure, when Siemens developed the Vidoson which
was mainly suitable for gynecological and abdominal examination"
(Hertz, private communication). The collaboration between Effert
and Siemens, initiated by Hertz, was to continue. Siemens's Vidoson
was brought onto the market in 1967.[24]

More than a decade had elapsed since Howry and Wild first
succeeded in generating images with reflected ultrasound. The
technology for producing these images was becoming widely available, and so in a sense Howry's vision of an all-purpose imaging
technique had been realized. Yet what was emerging differed quite
fundamentally from what he had had in mind. Howry had envisaged a technology that would complement existing x-ray tech-

niques in the hands of his fellow-radiologists. He had defined a space for ultrasound in terms of deficiencies in existing techniques, and much of his work had gone into trying to produce images of a *quality* to satisfy radiologists. The career of diagnostic ultrasound was not evolving in this way at all. Its development phase showed it being separately constituted within a number of medical specialties, each of which represented a distinctive set of clinical questions and a distinctive set of demands on the design of any commercial product. In some cases, as we saw, slight modification rendered an existing product useful enough. Faced with minimal entry barriers, manufacturers were more than willing first to lend, and then to sell, flaw detection equipment to ophthalmologists, cardiologists, and neurologists. Studied in relation to the clinical practice of gynecology and obstetrics, however, it became clear that were ultrasound to be of use major adaptation was needed. Out of the contacts between Donald and Kelvin Hughes grew a development program through which a true imaging device emerged, tailored in particular to abdominal examination. This entailed a growing commitment on both sides: of reputation on the one, of financial resources on the other. Socially, as well as technically, the constitution of ultrasound as a diagnostic tool differed substantially from one area of medical practice to another.

Diffusion and Accommodation, 1965–1980

Bringing ultrasonic imaging devices to the market had proved no easy matter. The long development phase through which the technology passed was the distant history of a variety of collaborations, development programs, and business histories, each of which reflects specific local circumstances. As we move into the evaluation and accommodation phase, in which the technology's medical and commercial position would have to be consolidated if it was to survive, it is becoming clear that obstetrics, as much as radiology, was to be a major context of use. Here we directly confront some of the most complex issues raised in chapter 2. On what basis, and by whom, would the technique's uses in abdominal examination be assessed? Will we find the kind of relationship between structural issues (such as control of the technology, adaptation of practice, and market structures) and problematizations which the analysis of chapter 2 leads us to expect?

There are clear indications of the general importance that this area of examination was assuming. We saw how Siemens, one of the giants of the x-ray industry, had nevertheless designed its Vidoson principally for obstetrical and gynecological application. Radiological attempts at taking control of diagnostic ultrasound were crumbling, and a quite different modus vivendi was emerging. We can see it in the research program initiated by Howry in Denver, Colorado, which had been the very basis of radiological ultrasound. In Denver collaboration with the university's department of obstetrics and gynecology was becoming central. By 1965 studies were being carried out with the contact scanner completed in 1962 (this seems to have been the basis of the first commercial instrument produced in the United States).[25]

Neither on the industrial side nor on the medical side were matters smooth sailing in the late sixties. In some cases technical difficulties were compounded by organizational or managerial ones. The Kelvin Hughes/Smiths Industries project is a case in point.

In 1967 the development work was completed and production of the five models that had been ordered was about to begin when, as former Smiths engineer Brian Fraser put it, "the axe fell." Strategic considerations, intended to bring about a rationalization of the firm's activities, led the Smiths main board to plan disposal of the Glasgow-based ultrasound work. This included the medical diagnostics work, and also the larger work in industrial testing, which were to be sold off to a Swedish firm. In the course of these negotiations, the diagnostic ultrasound work was eventually separated out and sold separately to the Scottish-based firm Nuclear Enterprises, which manufactured, among other products, gamma cameras for radiodiagnostics. The medical project, with both Brown and Fraser, moved to Nuclear Enterprises.

The subsequent period was a very difficult one. Although some thirty instruments were sold, major production had still not begun. The problem, as Fraser explained, was that electronic technology was advancing so rapidly. The Nuclear Enterprises group recognized that an instrument based on available technology would very rapidly be rendered obsolete, while the new solid-state technology did not seem quite ripe for exploitation. The activity was far from profitable; continuing development was an act of faith on the part of Nuclear Enterprises management. It was only in January 1972 that the device, now called the Diasonograph, finally came onto the market. It was sixteen years since Brown and Donald had begun working together—a vivid example of the effect of external constraints on even the most promising development process.

Difficulties were not all on the industrial side. There was also resistance among the medical profession, with not all obstetricians immediately persuaded of the claims being made for the technique. The discussant to a paper presented by the Howry and Holmes group in 1965, a Dr. Kiekhoffer, expressed what many must have thought. He did "not think it likely that the echo technique will replace either trial of labor, the agonizing reappraisal of clinical pelvimetry, or that indefinable entity clinical judgement, any more than did the advent of x-ray pelvimetry." The data, for example on normal fetal growth in utero are "most interesting," to be sure, but the echo technique "is unlikely to render obsolete the intellect of the obstetrician in making the final choice as to abdominal or vaginal delivery!" We can see here a possible source of obstetricians' unwillingness to grant the technique the significance its protagonists were demanding. If clinical decision making came to be dictated by this new machine then there was a danger of the routinization of the obstetrician's tasks. Moreover, if obstetricians were not universally enthusiastic, neither were radiologists prepared wholly to surrender the technique to the various treating specialties.[26]

Despite a degree of medical resistance, and despite the difficulties into which manufacturers could be led, ultrasonic imaging was gradually making its way in the hospital. From the end of the 1960s we can see this reflected on the one hand in industrial involvement and on the other in clinical activity.

A growing number of manufacturers were beginning to offer ultrasonic equipment specifically for medical applications. A survey by Arthur D. Little, commissioned by the National Science Foundation in the 1970s, suggested that there were five firms selling on the U.S. market by 1969 (including one Japanese firm, Toshiba) and twelve by 1973. (By the end of the 1970s there were perhaps thirty-seven firms offering equipment worldwide.) Moreover, just as the development phase had shown the technology being constituted within a variety of clinical practices, with only partial overlap in the research done (e.g., common research on safety), so too with markets. In place of a single market, e.g., within radiology, a set of partially overlapping markets was emerging. This gave the industry a particular structure. The Arthur D. Little review referred to it as follows:

Throughout their development, ultrasonic imaging devices have tended to be specialized for particular medical applications. It is partly . . . a consequence of the special characteristics needed in each medical speciality. Ophthalmology is confined to relatively small volume; cardiology requires a snapshot (a scan completed in 1/15 of a second or less) . . . obstetrics requires a relatively large imaged volume; etc. While the instruments offered by various manufacturers have diverged to meet these special needs so that not every company in the field of ultrasonic imaging is a direct competitor of every other, they have all built on a common base of knowledge and experience. (Colton 1982, section V)

Correspondingly, within each of the specialist areas, consensus gradually emerged as to what the clinical uses were to be, as we would expect. The processes by which consensus was reached differed from one specialty to another. In both neurology and ophthalmology there was long dispute over the relative roles of A-scan and B-scan instruments.[27] In relation to obstetrical and gynecological applications a particularly buoyant market was emerging, with rapid expansion in use. This is to some extent to be understood in terms of a gradual displacement of x-rays, by this time acknowledged to be unsafe, from obstetrical practice.[28] Obstetrical radiology (using diagnostic x-rays) was of course the province of the radiologist. Diagnostic ultrasound, if it could do the same job better (for example, more safely), might provide a means of reducing the professional dependence of obstetricians on radiologists. We see that there are significant interests at stake here: not only radiological hegemony in the imaging field but also the (possible) interest of obstetricians in rendering their own practice more independent. How is this situation reflected—or resolved—in the problematizations through which ultrasound's future career would be made? When we turn to the research through which ultrasound's applications in the abdominal area were established and its utility assessed, we shall see that the professional concerns of obstetricians, not radiologists, soon dominated.

We can see the connection in the development of the research program initiated by Howry. As concerns relating to obstetrical and gynecological uses of ultrasound came to shape the work in Denver, a clear shift in the kinds of research questions posed followed.

Where previously central questions had reflected radiologists' concern with morphology—e.g., demonstrating a "correlation between somagrams and gross anatomic detail"—they gradually came to reflect obstetricians' interests in "process" and mensuration. We can look specifically at the interaction between obstetrical concerns and research through following the work in Glasgow further.

By 1965 ultrasonic measurement of the biparietal diameter of the fetal skull could be presented as a vital means both of assessing fetal weight and of assessing the effects of complications of pregnancy upon fetal growth. Persuasion of the medical community would be facilitated if the relationships could be presented in a way that seemed to suggest easy assimilation into practice. "A simple rule" was deduced: "If the ultrasonic measurement . . . is 8.5 cms or more the baby is unlikely to weigh less than 4 pounds" (Willocks et al. 1964).

As well as a gradual move in the direction of measurement, studies also show an attempt to push use of the technique back into earlier stages of the pregnancy. "How early in the pregnancy can the fetus be visualized?" The question is related to another question, "How early in pregnancy does it make sense to carry out, or to require, an ultrasonic examination?" As the applicability of the technique was pushed back—from fourteen weeks in early work to nine weeks by 1963—advantages over other available techniques could be given further emphasis. This reasoning could appeal to the obstetrical community at various levels of interest: "Foetal echoes are detectable several weeks before the skeleton is visible by x rays. Results of the sonar examination are obtainable immediately which is an advantage over the time-lag experienced with biological tests for pregnancy."

Thus not only can the advantage over x-ray be articulated, but the technology also serves obstetricians' professional commitment to regular prenatal examination beginning early in pregnancy. Ultrasonic scanning early in pregnancy both facilitated and was legitimated by early prenatal examination (Hiddinga and Blume, forthcoming). Further work was to see application of the technique pushed back still further into pregnancy, with Willocks's simple rule of thumb of 1964 replaced by increasingly complicated decision algorithms (Campbell and Newman 1971).

Obstetric ultrasound had been retained in the province of the obstetrician, and the way was clear for large-scale adaptation of practice. In many leading departments ultrasound scanning was becoming a common element in prenatal care, with cephalometry, often done on a serial basis (to check development), the major application. (Donald reported that in 1972 it accounted for some 60 percent of the more than 4000 examinations made by this technique in his department (Donald 1974).)

If all pregnant women were routinely to receive an ultrasound scan in their prenatal care (as some obstetricians were urging by the early 1970s: see Donald 1974; Oakley 1984, 164–171)[29] the commercial attractiveness of obstetrical ultrasound is scarcely surprising. This stimulus to further development work meant that various sorts of ultrasonic devices came to be offered to obstetricians. We have seen how the Nuclear Enterprises Diasonograph was put onto the market in 1972, essentially for obstetrical applications. In parallel with all this, there emerged also a fetal heart detector, and then a fetal monitor, based on Doppler ultrasound. Sonicaid, a Sussex-based firm, started to market a system to monitor the fetal heart in 1971. Instead of an image, this produced a heart trace on an oscilloscope or chart. It sold for around £1000 ($3500), compared to the Diasonograph at £25,000 to £30,000 ($80,000 to $100,000).[30]

Success could bring problems, as Nuclear Enterprises discovered. By 1977 they were producing some fifteen Diasonographs per month, and a backlog of seven months of orders had built up. The directors of the firm saw the prospect of ultrasound coming to dominate Nuclear Enterprises, changing the character of the firm. Unwilling to let this happen, they decided to sell off the ultrasound division.[31] EMI, already a major shareholder in the firm, decided to take over the ultrasound work. This was 1977, the heyday of the CT market. Competition had forced EMI to tie up all available resources in further development of the CT scanner, and it was itself running into problems with deliveries. It is a measure of the prospects ultrasound was seen as having that EMI should have chosen this technique as a means of consolidating their place in the medical field. EMI thought that the existing Nuclear Enterprises product would complement the much more advanced system that they planned to develop. This project was never to be.

From the 1970s into the 1980s, both commercially and scientifically the buoyancy of ultrasound was maintained. Diagnostic ultrasound had not been effectively incorporated into the interorganizational structure centered on x-ray, and control of the technology remained fragmented. Though this fragmentation had made for a protracted and uncertain development process, it now contributed to ultrasound's commercial success. Because of it an individual hospital might well purchase more than one ultrasound unit, to be located in different departments. One study of changing availability of the technique in the United States found that between 1975 and 1980 the mean number of machines (in thirty-two leading hospitals) rose from 1.6 to 6.1 per hospital (Janus and Janus 1981). These were not located in radiology departments alone—77 percent of cardiology departments, 66 percent of ophthalmology departments, and 58 percent of obstetrics departments had their own ultrasound units. Of the thirty-two institutions surveyed, twenty-nine had plans to purchase one or two additional machines in the course of the following year. "Despite the fact that the number of centers providing ultrasonic diagnostic services is doubling every two years," reported a leading figure in medical ultrasound, "most centers report that the number of patients they examine also doubles every two years" (Kossoff 1978). This growth in utilization is reflected in industrial activity. In Japan, where the market for ultrasound equipment grew at a remarkable 40 percent per annum, compared with 15 percent for medical instrumentation generally between 1976 and 1978 , many major manufacturers (such as Toshiba and Aloka) made this a priority field for R&D work (DGRST 1980).

The Feedback Phase: 1969 to the Present

The notion of a feedback phase derives most obviously from economic theorizing. Attempts at improving commercially available instruments (the "types" actually available for use, in Hård's terminology) might be initiated by manufacturers where market potential seems to justify it. Thus, I have suggested that manufacturers felt it worth trying to develop new ultrasonic devices for abdominal examination because of the growth in these applications. Industrial surveys confirm that firms have behaved in line with expectation: "Because the ultrasound imaging market is growing more rapidly than any of the other diagnostic imaging markets, expenditures for research and development in ultrasound are being increased substantially by manufacturers. However, most of these expenditures are for improvement of current products and not for the development of advanced systems" (Hamilton 1982, 83). This study indicates a pattern of continuous marginal improvement that has been suggested to be typical of the medical instrumenta-

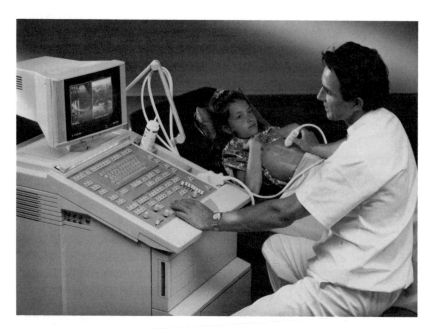

Figure 3.10
Modern ultrasonic device in use. Note alternative transducer heads available.
Reprinted by permission of Philips Medical Systems.

tion industry in general (Hartley and Hutton n.d.; see figure 3.10).
The conceptual scheme set out in chapter 2 suggests that we pose some less familiar questions—that we look not only at the *volume* of investment in improvements to the technology but at the structure and content of the work too. There are two reasons for this. R&D aimed at improving available diagnostic ultrasound instrumentation has been conducted not only by manufacturers but by users and other researchers as well. Economic incentives do not wholly explain what goes on in the feedback phase. A successful technology provides a basis for research on which more reputations than one can be built, just as much as it provides a basis for profit. Indeed it is perfectly possible that a large and extended reputational structure—with international journals, scientific associations and the rest—becomes established around such a technology. That is what has happened here. With half a dozen international journals devoted to medical ultrasound there is plenty of scope for publication, and papers continue to stream out. The second reason is to do with structures of control. In this chapter we have established what

has been described as a set of "partially overlapping markets." Many firms produce instruments for more than one of these markets, but not all firms are competing with all others. At the same time, and corresponding to this, we have a set of specialist practices into which ultrasound has been integrated. What one might expect is a complex feedback process in which both the technology as such and its specialized applications outside radiology provide bases for investigation. The result ought to be on the one hand developments affecting the technology more or less across the board and on the other further articulation of its various specialist applications. I want to show that this is indeed the case by turning finally to look at the evolving work of a research group that has been central to the field almost from the beginning.

In 1959 an ultrasonics research section had been established in the then Commonwealth Acoustic Laboratory, an establishment of the Australian health ministry. George Kossoff, a physicist, was placed in charge of the section, which was intended to be a general center of technical expertise in the ultrasonics field. From the outset research was pursued through a network of close cooperative relationships with medical practitioners in a variety of specialties (Kossoff 1975). At the same time, the group had always been concerned with the *physics* of its work in a way that had not been true of most other groups we have looked at. This is not surprising, since the core group was composed of physicists and engineers, but it has important implications.

Obstetrics was an area of interest almost from the outset, and by mid 1962 the first obstetrical echograms were being recorded. Other areas were added. Two that were introduced into the research program in 1964, and which were to remain of importance, dealt with breast and cardiographic studies. Both were to prove problematic. "[B]ecause the ultrasonic appearance of a normal breast is highly variable" this organ proved highly recalcitrant to ultrasonic investigations. Though the female breast retained its allure, and its challenge, progress would be made only some years later, following a major technical breakthrough. The cardiographic work yielded practical results sooner, and "the CAL ultrasonic cardioscope" was developed specifically for cardiological purposes (Kossoff and Wilcken 1967).

Cardiographic studies, breast studies, and obstetric studies were proceeding in parallel, on the basis of collaborations with specialists in the various fields. But Kossoff was also exploring the possibilities of the technology in a quite different and more general way, seeking a means of obtaining more *information* from the ultrasonogram. This is quintessentially a physicist's question.[32] What Kossoff discovered was that by substituting another means of recording the data for the kind of cathode ray tube then generally used, more information could be obtained. The new technique, which came to be known as gray scale scanning, replaced the storage-tube oscilloscope with a conventional oscilloscope and a film (Kossoff 1972). It took advantage of the fact that a 35 mm film can record at least ten shades of gray (corresponding to the size of the echo), "greatly increasing the diagnostic information content." With this technique, it was possible not only to record accurately the size and position of an organ but also to investigate its internal structure. All the instruments being used by the Sydney group were modified to work in the gray scale mode. The gray scale technique was to prove of relevance to virtually all applications of ultrasonic imaging. The next thing was to see what could be done in areas in which ultrasonography was already being used as well as in areas that had thus far defeated it. General technological advance is the stimulus to further articulation of specialist applications. Success in the early diagnosis of breast disease would offer a major new market to ultrasound and provide a major challenge for radiology. In the early seventies Kossoff returned to the breast problem.

The female breast is a complex structure made up, in proportions that change roughly with age, of glandular and fat tissue. Was it going to be possible, by measuring the velocity of ultrasound, to "(1) characterize the tissues present in the breast, and (2) show the range of variation of these tissues, both within each age group and across the age groups"? An in vitro study of excised tissue provided velocity measurements for the various components (fat, muscle, etc). A study of "an allegedly normal pre- and postmenopausal population of subjects in the United States and in Australia" (carried out together with E. Fry of the the University of Indiana) showed an "average velocity in the premenopausal group" of 1510 meters/sec, though "large deviations from this average are common." The postmenopausal group was lower, on average (1468 meters/sec) with, again, wide variation (Kossoff, Fry, and Jellins 1973). But what did these figures mean? Could they be made to mean something about the structure of the breast? That required a further study.

For this purpose "a study was undertaken on 19 normal subjects (ranging in age from 19 to 63), for whom x-ray films of the breast (mammograms) were available, in order to determine possible correlations between acoustic velocity measurements and the primary type of tissue present in the breast." The hope was that by this means the structure of the breast—as defined by x-ray—could be shown adequately to be given by ultrasonic measurements. The study seemed to work; both the *variability* in young subjects and the effects of menopause were shown to be "a function of the relative amount of adipose [fat] and glandular tissue." But did all this have any diagnostic significance? The work was published not in a medical journal but in the *Journal of the Acoustical Society of America*, and it offered no more than hesitant suggestions regarding diagnostic relevance (Kossoff, Fry, and Jellins 1973).

When Kossoff returned to making breast echograms using the new gray scale technique, the velocity measurements were useful in setting the apparatus for the patient being examined. But to find a criterion by means of which he could decide if pathology was present, he turned to something much simpler: "Although the echo pattern of the breast varies considerably between patients, structural similarities exist in the breasts of any individual. Small abnormalities can be most easily differentiated by comparing the echograms of the pathological breast with echograms of the normal breast." (Jellins et al. 1973/74)

For the physicist, writing in a physics journal, normality is defined in terms of a distribution of values (velocity measurements). By contrast, it seems as though for a medical public there is more to be said for a very different criterion deriving from the comparison of one breast against the other, *which is called normal.* It was in fact only in 1976 that Kossoff felt able to claim a role for ultrasound in the diagnosis of breast cancer, and then only a modest supporting one. In chapter 4 similarly contrasting conceptions of normality will reappear, and we shall explore in detail the sorts of difficulties to which the distinction gives rise. In that case, as we shall see, they were to prove fundamental to the career of the technique of infrared thermographic imaging.

By the late 1970s there was a good deal of discussion as to how much more sophistication should be built into ultrasound devices. Reducing operator dependence through interfacing with computers was technically possible by then, but did physicians want it? Specialization was central to the career of diagnostic ultrasound, as we have seen. Alternative conceptions of what exactly the technique should be *for,* present from the start, had all taken root. Specialization in use had been associated with the formation of "partially overlapping markets" and, as a result, specialization in equipment. But how far could this be taken before it became unprofitable for manufacturers?[33] The fact that radiologists had never been able to control either use or development of diagnostic ultrasound had a number of consequences. Fragmentation in control corresponds to a segmented market, in which manufacturers can avoid competition—but at the price of potentially small market segments.[34]

Market growth, in the case of a medical technology, is likely to correspond to growing recourse to the technology in processes of diagnosis or therapy. We have seen that this has indeed been the case with ultrasonic imaging. In obstetrics, in particular, use of the technology has become more or less routine. In her book *The Captured Womb,* Ann Oakley writes, "There is no doubt,' confirmed Stuart Campbell and D. J. Little, 'that the development of the real-time scanner has transformed prenatal care. Ultrasound is now no longer a diagnostic test applied to a few pregnancies regarded on clinical grounds as being at risk. *It . . . should be regarded as an integral part of prenatal care.*" (Oakley 1984, p. 165, italics added).

Precisely the routine character of obstetric ultrasound has been the basis for recent criticism from the women's movement, concerned with the way in which technology has impacted on childbirth. Critics have called into question not only the manner in which use has become routine, no longer in need of justification, but also the safety of the technology.

The number of studies on the biological effects of ultrasound has begun to rise sharply in the past few years, with fetal development among the most intensively investigated areas (Stewart 1985). Earlier data now prove to be surprisingly limited. Epidemiological studies suggest that there could be some effects on humans associated with exposure in utero, and there is the possibility of genetic effects. Was safety ignored in the development of diagnostic ultrasound? Were obstetricians and others unconcerned with whether or not the technique was safe? They were certainly not unconcerned, and probably concern over safety was a factor in the initially slow acceptance of ultrasound in obstetrics. Research into its safety was done, and in a 1974 article in the *American Journal of Obstetrics and Gynecology* Professor Donald discussed it at length (Donald 1974). However, not only was the research limited compared to the amount of work done on the effectiveness with which ultrasound diagnosed, distinguished, and measured, but it is noteworthy that most of the important research was done only in the early 1970s. By then significant investments had already been made; ultrasound was entering widespread (if not yet routine) use.

The success of ultrasound stimulated continuous searching for ultrasonic applications and techniques, both by industry and by groups such as Kossoff's. This has now gone on for twenty years. The very considerable effort has resulted in a stream of improved and diversified techniques. The career of diagnostic ultrasound, hitherto, seems to attest to the utility of our conceptual scheme. We have found the technology constituted within a complex social structure, and we have found the marks of this structure in the work of innovation. As for whether recent protests, and new research attention for the safety of ultrasound, find any reflection in patterns of use, only time will tell. Should current concerns significantly affect the future of the technology, we would be obliged to refine our model. We would have to note, first, that feedback processes can *change* in the course of time, provoking changes in the fundamental criteria governing the innovation process. This is perhaps not so surprising, in that the very idea of government *regulation* is intended to achieve just that. But the further implication of the ultrasound example, were current concerns to prove of significance, is to suggest that the ultimate consumers of medical technology can, under certain circumstances, influence its further development.

4

The Constitution of Thermal Imaging

Introduction

As with ultrasound, modern techniques of thermal imaging, or infrared thermography, derive from physical principles worked out in the nineteenth century. More immediately, and again like ultrasound, infrared thermography involves the redeployment into medicine of technology initially developed for military purposes. However the conditions under which this military technology was developed were different, and so were the circumstances under which its possible relevance for medical imaging first arose. Military research on ultrasound began during the First World War. By the 1930s, when its possible applications in medicine were first suggested, there were nearly 20 years of experience in Britain, France, Germany, and the United States. By the end of World War II expertise was widely available in both Britain and the United States and the postwar development of ultrasonic imaging could profit from the demobilization of technical expertise, as well as from an interest in industry in redeploying its technology to peace time ends. The situation in the 1950s was very different. With the Cold War at its height, military technology was now being developed under conditions of obsessive secrecy (see Shils 1956). Expertise in the area of infrared technology, in Britain principally limited to military research establishments, was not easily accessible. That the possibility of its application to medical imaging arose at all is largely due to a small number of physicians possessed of a clear vision of the diagnostic task confronting them. That that task involved early detection of cancer enabled them readily to win support for their quest. As we follow the subsequent attempts to develop an effective diagnostic device, we shall find that what was entailed, both for physicians and for industrialists, differed profoundly from the ultrasound case.

In 1800 Sir William Herschel had shown that the electromagnetic spectrum passes beyond the red light of the visible spectrum into a region where greater heating takes place than in the visible part. This region of the spectrum he called the "infrared." Forty years later, his son Sir John Herschel was able to make this infrared spectrum visible in the form of crude pictures, which he called "thermographs." This he did by coating paper with lamp black and soaking it in alcohol. On exposing the soaked paper to infrared radiation, alcohol evaporated in proportion to the radiation received, thus rendering the paper lighter in color. In the course of the nineteenth century a variety of devices were developed that enabled this region to be investigated more sensitively. These devices for measuring thermal radiation included Nobili's thermopile (a series of pairs of dissimilar metals, or thermocouples, which is sensitive to temperature differences) in 1829, and the bolometer (a resistance thermometer connected to a galvanometer, developed by Langley in 1881). In the latter part of the century significant theoretical advances in physics led to a quantitative understanding of this thermal radiation. It became known that any object above absolute zero (zero degrees Kelvin) emits energy from its surface as a spectrum of different wavelengths whose form depends both on its temperature and on its character or "emissivity." In 1884 Boltzmann, building on earlier work by Stefan, formulated what we now call the Stefan-Boltzmann law, according to which the total radiation from the surface of a "perfect emitter," or "black body" is proportional to the fourth power of its absolute temperature. A few years later the German physicist Wien formulated the law that bears his name, according to which the shape of the spectrum projected by the perfectly emitting black body is determined by its temperature. In particular, the wavelength at which most radiation is emitted varies inversely with the temperature—the higher the temperature, the lower the maximum wavelength.

In the decades following, interest in infrared developed in physics, in chemistry, and in physiology. Physicists were at first interested largely in relation to investigations of the electromagnetic spectrum and then, together with chemists, as a means of investigating molecular structure. By the 1940s infrared absorption spectrometry gained great importance for the characterization of the structures of organic compounds and had found considerable application in the chemical and petroleum industries (Rabkin 1987). Visualization of the infrared, Sir John Herschel's "thermography," lacked the same importance. In 1929 the German physicist Czerny had developed a more sophisticated technique, based on the same principle, and announced it in *Zeitschrift für Physik*. But it was only years later, in the post World War II period, that a growing interest in the military possibilities of infrared detection and visualization led to any significant volume of further research in this area.

Meanwhile work in the 1920s had shown that it was possible to measure the radiation from the surface of the human body and had suggested certain interesting features about the pattern of this radiation. It seemed in the first place from Cobet and Bramigk's work (1924) that the human body behaved effectively as a "black body." That meant that Wien's law applied, so that the wavelength at which most radiation was emitted should be related to the temperature of the emitter. Moreover, the Stefan-Boltzmann law should also apply, so that the total radiation should be proportional to the fourth power of the absolute temperature. Bohnenkamp and Ernst (1931) found that this law did apply and that there were ways of using the emission of infrared radiation to measure surface (i.e., skin) temperature. These suggestions in the literature were picked up and developed further in an important program of research in the early 1930s.

In 1934 J. D. Hardy, a physiologist working at the Russell Sage Institute of Pathology in New York, published the first of a long series of articles on the radiation of heat from the human body (Hardy 1934a). Hardy's fundamental interests were in thermal aspects of metabolic processes and the means by which the body maintained its thermal equilibrium. It was from this conceptual framework that his interest in skin temperature derived. "In the past few years," he wrote, "the interest in an accurate knowledge of the temperature of the skin and of the energy radiated therefrom has greatly increased. This interest has been due on the one hand to the important part played by radiation in the theories of ventilation and on the other to desire to make a more accurate analysis of the factors involved in human metabolism."

In order to investigate metabolic processes through the study of skin temperature, the first thing needed was an accurate measuring instrument. Hardy and his colleagues developed a radiometric device for the measurement of skin temperature. Subsequently they attempted to demonstrate the superiority of their new device and to substantiate the view that the human body functions as a black body. Recall that it was this supposition that justified use of the radiation device for temperature measurement. Their research program, though focused in physiology (most of the work was published in the *American Journal of Physiology*), also took up a number of clinical issues.

During the 1930s and 1940s a variety of clinical interests in skin temperature measurement also emerged. Skin temperature was significant in relation to circulatory disorders, rheumatoid arthritis and various psychiatric disorders as well as to interventions such as anesthesia and the effect of vasodilatory drugs on circulation. Peripheral vascular disease was probably the most investigated of these topics. From this clinical work emerged the idea of the thermal *symmetry* of the body. Linder and his colleagues, interested in the adaptive functioning of schizophrenics, had come to believe that this condition was associated with abnormal response to temperature change. Large numbers of skin temperature measurements were made on schizophrenics and controls at symmetrical locations on the body (Freeman, Linder, and Nickerson 1937). In order to see if they could make do with half as many measurements, they looked at differences between these sets of positions. Results demonstrated a "high degree of similarity between temperatures of symmetrically placed areas."

Radiation-measuring instruments of the sort Hardy had developed, studied, and used were rarely employed. Investigators stuck to contact devices: typically, a thermocouple placed in contact with the skin and giving its reading as an electrical current on a galvanometer.

Only after World War II did any real interest in instruments for detecting and measuring infrared radiation emerge, and that within a very different context. An infrared detector could locate heat-emitting objects that were camouflaged and thus not visible in ordinary light. If this radiation could be converted to a visible image, then there were intriguing possibilities of "seeing in the dark." Much of the existing expertise in infrared technology was deployed in this area. Companies such as Baird Associates, which had been making infrared spectrometers, developed an advanced version of Czerny's apparatus with support from the U.S. Air Force. Most of this research remained classified for many years.

As part of this research effort, a good deal of attention was being devoted to the search for improved means of detecting infrared radiation. A very early alternative to Herschel's lamp black and alcohol soaked paper was found in a phenomenon discovered by

Becquerel. In 1843 he had discovered that the phosphorescent afterglow that ultraviolet or visible light produces in certain chemical substances was quenched by subsequent exposure to infrared. A century later, this finding was used as the basis of a method of making thermal pictures. Urbach, Nail, and Perlman applied a phosphorescent material to a surface to be investigated and illuminated with ultraviolet light. Temperature patterns in the surface could be demonstrated by the quenching of luminescence in proportion to the infrared radiation. This technique was given the name contact thermography.

By the end of the 1950s there were two distinguishable sorts of infrared detector: one based on materials sensitive to temperature (thermistors) , the other on materials sensitive to light. Photosensitive materials (such as indium antimonide) were the subject of a good deal of research, including at the Royal Radar Establishment at Malvern in England. These two sorts of detector had rather different properties. Thermistors, temperature-sensitive electrical resistors made of metal oxides, responded much more slowly to temperature change than did photosensitive materials. On the other hand, it was known that photosensitive detectors worked better if they were cooled to very low (e.g., liquid nitrogen) temperature. Such cooling was not needed with the thermistor. Infrared devices of these kinds, developed for military purposes, began to be declassified in the late 1950s. The Baird Associates' Evapograph was made public in 1956 and the Barnes thermal camera in 1957 (a description was later published in the *Journal of the Optical Society of America*: see Astheimer and Wormser (1959)).

R. Bowling Barnes, a onetime Princeton infrared physicist who established the Barnes Engineering Company, wrote of the interest in infrared which developed as follows:[1]

During the past quarter century, interest in the field of infrared has increased tremendously. The capabilities of several infrared instruments have been widely recognized by industry, and today infrared spectrometers, for example, are being employed routinely in most analytical chemistry laboratories. Since all objects radiate infrared and since this emission may be utilized without the need for any active illuminating process, infrared has contributed materially to the solution of many problems connected with national defense and the exploration of space. Among the developments to which it has contributed are the Sniperscope; the heat-seeking head of the Sidewinder missile; the horizon sensors used in the stabilization of spacecraft, such as that of the Project Mercury program; the temperature-measuring radiometers of the Tiros weather satellites, and the recent Venus probe, Mariner II. (Barnes 1963)

Thermal Imaging: The Exploratory Phase, 1955–1963

In the mid 1950s, then, expertise in advanced infrared technology was located not in the medical imaging industry but in military R&D establishments and among industrial defense contractors. It was to be sought out, redeployed, not by radiologists interested in the precise representation of morphology but through a specific need for accurate temperature measurement. For industry the technique seemed to offer a number of companies that did not make x-ray equipment the possibility of gaining access to the developing medical imaging market.

In 1955 in the course of some studies of drug-induced tumor pain, Ray Lawson, a Montreal surgeon, discovered that breast tumors seemed to be associated with an increase in skin temperature of some two degrees over the affected part. In the course of the following year Lawson began to investigate the clinical significance of this finding, both as a possible means of diagnosis and as a possible means of following the progress of therapy. In August 1956 Lawson published his first results in the *Journal of the Canadian Medical Association.*

In this work Lawson used a contact thermometer, a thermocouple, just as other medical investigators had been doing since the 1930s. The diagnosis—presence or absence of a tumor—was based on a temperature *difference* between symmetrical points on the two breasts. The underlying physiology, the reasons for this apparent temperature elevation, seemed relatively unproblematic. It was not difficult to imagine the rapid cell division characteristic of tumor growth being associated with accelerated local metabolism. Far more problematic was the technique to be used to measure these differences with the desirable degree of accuracy. After all, Lawson was giving "temperature" a quite different diagnostic significance than it had traditionally had in medicine. A difference of only one or two degrees could imply radically different diagnoses. How could the vital measurements best be made? It was in searching for the most sophisticated temperature measuring instrumentation he could find that Lawson was led to infrared technology. The timing was right. Although the technology had been developing under conditions of military secrecy, in February 1956 Lawson obtained a Baird Associates Evapograph, which was just being declassified. He soon decided that the imaging principle of the Evapograph rendered it unsuitable for his needs (Lawson 1957). In 1957 Lawson secured limited access to the device made by the Barnes Engineering Company. Unlike the Evapograph, this was a scanning device that built up an image through a process of line-by-line scanning, much as a TV camera. Here too there seemed to be problems. The Barnes instrument made use of a slowly responding thermistor, and Lawson was concerned that "the minimum scan time of approximately 10 minutes would preclude any practical medical use." Lawson carried on looking for an infrared device that would meet his needs, and in the subsequent years was able to make use of a number of instruments and to collect data from large clinical series of breast lesions.

The thermometric criterion for the diagnosis of breast cancer was original and important, Lawson argued, because it was based on a novel way of *conceiving* tumors, very different from the morphological characteristics at the heart of radiological diagnosis. He stated that "Unlike anatomical alterations that represent the cornerstone of breast cancer diagnosis, increased heat is a functional dynamic abnormality. A lump in the breast is defined as an area of increased density as compared to the surrounding anatomical structures. The presence of a lump is elicited by finger palpation, and it is usually impossible to detect one less than one cm in size . . . a clinically palpable lump is an inadequate and late sign of breast cancer" (Lawson 1964).

Though Lawson began to publish the results of this work in 1956, it was some time before it had any impact. In 1959 Kenneth Lloyd Williams, a surgeon at London's Middlesex Hospital, was interested in locating incompetent perforating veins in patients with varicose veins. It had occurred to him that the warmer blood in such a vein would be pumped to the surface of the skin, so that "the skin temperature might be greater at the point where a perforating vein 'surfaced,' and that the infrared emission from the skin would therefore be increased" (Lloyd Williams, and Handley 1960). Lloyd Williams hit upon the idea of trying to measure skin temperature through infrared radiation because it seemed to offer a number of specific advantages over a contact thermometer. For example, one could look at a small area in place of a point, which would correct for local variations. As a means of locating perforating veins, the method was not initially successful. But instead of pursuing his original problem by other means, which would perhaps have been the usual thing to do, Lloyd Williams used his (negative) results as "a starting point for a more general consideration of infrared emission from the skin under normal conditions and in pathological states" (see figure 4.1). Having thus determined on a shift in the focus of his work, Lloyd Williams soon became aware of Lawson's results, which were very relevant within the context of his new interests. The Middlesex surgeons turned their attention to the investigation of breast tumors.

Using a device like that demonstrated by Hardy, Lloyd Williams set about examining breast disease by a manual measuring-and-plotting procedure. Like Lawson, he took a temperature difference between symmetrically placed points on the two breasts, as his diagnostic criterion (Lloyd Williams, Lloyd Williams and Handley 1961). In 1961 a clinical series of 100 patients with a lump in one breast (thus with an existing diagnosis) admitted to their hospital in a 12-month period was published in the *Lancet*. Like Lawson, Lloyd Williams concluded that the "skin overlying carcinomata and inflammatory lesions was hotter than the mirror image area on the normal side . . . the skin over degenerative lesions was cool. Such findings are sufficiently constant to be useful in diagnosis."

Encouraged by these results, Lloyd Williams decided to go further. Having visited Lawson in Montreal, Lloyd Williams concluded that he too would need a scanning instrument, like the ones Lawson had been able to borrow. Back in England he came into contact with an engineer, C. Maxwell Cade. Cade, who worked for Smiths Industries, had been trying to develop a thermal scanning device for detecting ships in fog.[2] However it had not been possible to identify a market, and his project had been shelved. It looked as though there might be a common interest: Lloyd Williams wanted an infrared scanning device, and Cade was interested in establishing the possible utility of the device he had built. In the course of 1961, Cade's prototype was adapted to look at the human breast. In their 1961 *Lancet* paper, Lloyd Williams and his colleagues added their first tentative experiences with this new instrument. In a matter of minutes it had generated a "heat picture," and although not very sensitive (temperature differences had to be more than 2 degrees to be visible to the eye in these pictures), "crude though it is, it does show that heat-scanning of an area of the body is

Figure 4.1
Photographs and thermographs of members of four races, showing that infrared radiation is a function of temperature alone. Visible differences result from different amounts of insulating fat. From Barnes 1963, 873. Copyright 1963 by the AAAS.

practicable; and improvements in scanning speed and detector sensitivity could produce a new tool with great potentialities in clinical medicine and research."

Results were encouraging. There seemed good reason to think that not only could precise skin temperature measurement provide a diagnostic method for early cancer but the pictorial scanning technology could also, in principle, provide precisely this accuracy. The next step was clear. Lloyd Williams would have to set about trying to stimulate industrial interest in building an infrared scanner adapted for medical use.

Meanwhile his results, published in a leading medical journal, had attracted the attention of other specialists concerned with detection and diagnosis of breast cancer. By the 1960s screening for breast cancer was becoming an important element in the practice of radiology, especially in the United States. Radiologists had developed a specialized x-ray technique, known as mammography, for this purpose. Among those who were intrigued by Lloyd Williams's paper was Jacob Gershon Cohen, director of the division of radiology of Philadelphia's Albert Einstein Medical Center, a man of great energy and entrepreneurial spirit. Gershon Cohen had first argued the importance of roentgenographic breast screening in 1937, and he had campaigned tirelessly for it since that time. After twenty-five years of experience with mammography, he read Lloyd Williams's 1961 paper in the *Lancet* with what we can imagine to have been a mixture of excitement and alarm. Here, perhaps, was a better way of looking for breast cancer. If that were so, then a burgeoning and important element of roentgenological practice was threatened. If thermography were to live up to this early promise, then it was essential that the technique be incorporated into the practice of radiology, rather than of any other specialty that could use it to bolster claims to authority in the breast cancer area. Gershon Cohen decided to investigate, and he too set about finding suitable equipment to pursue this work.

Like Lawson, both Lloyd Williams and Gershon Cohen were led to establish contacts with firms having expertise in the field of infrared technology. The establishment of these contacts, and the subsequent collaborations, are characteristic of this exploratory phase of thermography. The necessary expertise, as we have seen, was largely derived from military work, and the firms Lloyd Williams and Gershon Cohen approached had no experience of medical markets. They were, however, interested in the possibility of capitalizing on their existing expertise and were receptive to the advances made to them.

In 1962 Lloyd Williams tried to persuade the managing director of Smiths Industries that the firm should produce a medical thermograph.[3] His arguments were well received. Smiths wanted to move into medical markets, from its traditional base in marine and control instruments. Lloyd Williams argued that since most of their expertise was in the area of measurement, it made sense for them to move into medicine through measurement, capitalizing on their expertise. But additional knowledge would have to be obtained. The Smiths Industries people knew that, despite their background, they were far from having state-of-the-art technology in the infrared field. Expertise in infrared detection, it was known, resided largely with the Ministry of Defense and was classified. Lloyd Williams discussed the plan with his colleague Sir Charles Dodds, Professor of Biochemistry at the Middlesex Hospital Medical School and an important member of many government committees. Dodds asked for a short paper on what was proposed. Within a few days he had been able to obtain agreement from the Ministry of Defense, giving access to advanced infrared technology. Slowly the relevant interests were being assembled. Contact was made with D. W. Goodwin of the Ministry of Defense's Royal Radar Establishment at Malvern, a leading authority on infrared detection. With Goodwin's help it did not take the Smiths engineers long; before the end of 1962 four prototypes of the new Pyroscan had been built. Using a photosensitive indium antimonide detector at liquid nitrogen temperature, the Pyroscan could produce its image rapidly. Thanks to Lloyd Williams's conviction, power of persuasion, and results linking infrared measurement to cancer, the medical thermograph had become a physical reality.

It was a reality potentially threatening to established radiological and industrial interests. Not surprisingly, Gershon Cohen—also seeking out appropriate infrared equipment—was engaged in establishing a similar (though effectively competing) coalition. His approach could not be to the traditional suppliers of radiological equipment. Accordingly, "When inquiry of several knowledgeable researchers in the field of infrared sensing devices was made, with almost universal accord, the authors were referred to R B Barnes, of Stamford, Connecticut. To him we are grateful for making available the military equipment for our first thermographic observations" (Gershon Cohen et al. 1964).

With equipment loaned by Barnes Engineering, Gershon Cohen and his colleagues started work in 1962. Lawson had passed this way, too, but whereas Lawson had been dissatisfied with the slowness of the instrument he was given and moved on looking for others, Gershon Cohen remained. Just as in Britain, where a common interest emerged between Lloyd Williams, Cade, and Smiths management, so too here. Despite the disappointment of the first results, Gershon Cohen and Barnes joined forces. In a veritable stream of publications, separately and jointly, the two men strove with might and main to bring the new measuring technology before the medical profession. For Gershon Cohen the future of that part of radiological practice in which he had made his name was at stake.

To defend it he was obliged to forge a coalition with a firm outside the established x-ray industry.

For Barnes Engineering a market of considerable growth potential seemed to be opening up. In 1963 the company went rapidly ahead with modification of their military camera, specifically aiming at medical use. This new model had a number of refinements over the military device. It could be calibrated with a gray scale to give absolute temperatures, and the sensitivity could be adjusted (so that the total black-to-white range could correspond to a difference of 18°F in "low sensitivity" or 1.8°F in "high sensitivity"). Unlike the Smiths Pyroscan, the Barnes Thermograph used a thermistor detector, not the photosensitive detector that Smiths had learned about at Malvern. It was thereby an inherently slower instrument.

Whatever the professional interests in breast cancer diagnosis may have been, industrial interests were not limited to radiological practice. These were firms without established medical markets or loyalties, naturally interested in establishing as broad a field of application within medicine as they could. Prototypes of both the Pyroscan and the Thermograph were now placed for trials in a number of medical departments: in Britain these were in units dealing with burns, vascular disturbance, in a cancer hospital and in a hospital for rheumatic disease. Prototype Thermoscans were loaned to a wide range of departments, including neurology, obstetrics, cardiology, trauma, and rheumatology.

In December 1963 Barnes and Gershon Cohen chaired a congress on thermography in medicine, held under the auspices of the New York Academy of Sciences.[4] Those to whom Thermographs had been loaned were invited to present their preliminary findings, and Lawson, Cade, and Lloyd Williams attended too. In his lecture to the meeting, Barnes had an interesting argument: "When one considers the medical significance attached to even a single body temperature measurement, whether made orally or rectally, it is clear that a thermogram, which may contain as many as 60,000 separate temperature measurements, should be capable of providing a wealth of clinical information" (Barnes 1964).

Preliminary assessments presented at the conference were very mixed, from the highly enthusiastic (in the evaluation of wound damage and healing) to the highly skeptical (a neurologist). Gershon Cohen and his colleagues presented a paper on their own studies of breast cancer, which included a presentation of seven case reports.[5]

While it is true that "we do not know just what aspects of malignancy we are detecting by thermography," nevertheless, they argued, "the temperature record of a patient's illness is so important that to omit its use on a hospital chart is to jeopardize accreditation by our professional sanctioning authorities. . . . It is obvious that we doctors will not abandon the use of the oral thermometer, but does that mean that we can afford to neglect use of modern heat sensing techniques which are capable of measuring the surface temperature of Venus at a distance of some 23,000 miles with an accuracy of better than 0.5°C."

The cautious optimism of the British researchers, Lloyd Williams and Cade, was in striking contrast to the aggressively entrepreneurial style of the Americans. For both Cade and Lloyd Williams, while first experiences provided reasons for optimism, there was a long way to go. Lloyd Williams summarized: "We know that heat scanning is practicable; we now need to know if it is of any value when applied to clinical medicine. We need to know which areas of medicine are worth exploring by heat scanning with some prospect of success, what are the possible snags and misinterpretations of the technique and what modifications in equipment could make it more effective."

In this first, exploratory phase of the development of infrared thermography, the central research question concerned the apparent relationship between the presence of an underlying breast tumor and a small elevation of skin temperature. This empirical finding led to the search for the most sophisticated radiation-measuring devices. Seeking out these devices, the physicians involved were brought into contact with firms possessing (or able to acquire) the necessary expertise in infrared technology. Barriers to entry, costs, and technical difficulties in adapting existing products were small. Firms were receptive to the idea of deploying their expertise in a new market and interested in the possibilities of the medical devices market. Within a year the first prototypes specifically designed for medical imaging were available. Not all could view this state of affairs with unconcerned interest. Diagnosis of breast disease was becoming an important element in radiological practice, and any alternative diagnostic technique posed a potential threat. Moreover since the firms were not the established suppliers of x-ray equipment, the threat to established interests was all the greater.

There was an important element of exploration involved in introducing infrared measurement into medicine: exploration of the *possibilities* of the new device. This resulted from commercial interests in establishing the broadest possible medical market. These firms had no established ties with the radiological profession. Both Smiths in Britain and Barnes in the United States placed prototypes of their new devices in a variety of clinical settings. But commercial interests are not all. We must not underestimate the sheer fascination that exploring the possibilities of a new instrument can exert, which reflects cognitive and social (reputational) interests in science. The general importance of this phenomenon for *laboratory* science—the extent to which research programs may be built around instrumental possibilities—is becoming clear (see, for example, Galison 1988, Wise 1988). When Lloyd Williams abandoned his original problem of perforating veins, or when Donald set out to explore the possibilities of ultrasound for obstetrics and gynecology, they were behaving as scientists seem frequently to do. Nevertheless, the possibility of early detection of breast cancer had given the initial impetus to investigation of thermography, and it remained at the forefront of attention.

The Development Phase: 1963–1965

The dawn of 1964 seemed to provide for cautious optimism on the part of those few physicians and industrialists who had made some commitment to infrared thermography. There were no major difficulties in the way of adapting military technology to medical uses and—in striking contrast to diagnostic ultrasound—the development period was to prove both short and unproblematic. Since no external radiation was involved questions of safety did not arise. Though without prior experience in the medical devices market, both Barnes and Smiths had thermographs for sale within a very short time. Admittedly the scope of the instrument's utility was far from clear, but there seemed reason to look forward to a further consolidation of what had been achieved. Significantly, the principal American grouping of Barnes and Gershon Cohen seemed far more convinced of the *proven* value of the technique than did their British colleagues (Cade and Lloyd Williams). Though we cannot analyze the sources of this conviction, it is plausible to assume that it reflects not only the different entrepreneurial climate of the

United States compared to Britain but also the deep commitment of Gershon Cohen, as a radiologist, to new technology. Although not all physicians were persuaded that mammographic screening offered benefits over palpation (or indeed over self-examination) sufficient to justify the acknowledged risks of radiography, there was a widespread commitment to provision of the service. Mammography had become an important component of radiological practice. If thermography was to find its clinical place in breast examination then radiologists had a fundamental interest in controlling it and incorporating it into their practice. On the industrial side, it was relatively easy for firms with expertise gained from defense work to develop a medical thermograph. No monopolization by the existing x-ray industry would occur during the development phase; instead a few more companies with experience in defense electronics began to dabble in infrared thermography. By the end of the short development phase commercial models were available from Barnes Engineering (the Thermograph), Smiths Industries (the Pyroscan), and two Swedish companies: AGA (the Thermovision) and Bofors.

The overall scale of clinical interest was also expanding, albeit slowly. In 1965 Gros initiated a major program of investigation in Strasbourg; others began in Italy, in the Netherlands, and elsewhere. In 1965 the first instruments were imported into Japan, and in 1968 the first Japanese thermography conference took place, with 19 papers being read (Atsumi 1977). Within this evolving network, the central questions for research were becoming more diverse as manufacturers and their associates made the case for the technique on a wide front. New problematizations were emerging as the relevance of the technique for certain other clinical specialties, where further work seemed justified, received sustained attention. In striking contrast to the ultrasound case, however, there is no question here of independent and parallel constitution of infrared thermography in distinct specialist contexts.

Among the areas in which further work seemed to be justified by preliminary studies was the field of rheumatology. The association between arthritis, inflammation of the affected joint, and rise in temperature—the notion that inflamed joints are hot—has been recognized since the earliest days of medicine. "But no one had attempted to do anything with that," explained Dr. Francis Ring of the Royal National Hospital for Rheumatic Diseases in Bath, "other than place a hand on the joint and say 'yes, it feels hot, I think it's inflamed.'"[6] Dr. Ring, who is a physicist, was therefore extremely interested when he read an article by Maxwell Cade explaining the principle of heat imaging. His colleagues, the rheumatologists, were also intrigued, and Cade was invited to bring the prototype on which he was working with Lloyd Williams to Bath. This he did on a number of occasions. When the first four prototype Pyroscans were placed for evaluation in 1962, one was loaned to the Bath hospital. First results were most promising, and a Pyroscan was purchased.

In other specialties thermography was less enthusiastically received. A 1965 paper by Gershon Cohen argued the utility of the technique for obstetrics and gynecology, especially for placental localization ("now possible in obstetrics without exposing mother or fetus to the risk of irradiation or radioactive isotopes"). But obstetricians were largely not impressed, and the technique made little headway there. Indeed Charles Gros, Professor of Radiology in Strasbourg, who played a major role in early investigation of the technique, himself soon doubted its applicability to abdominal studies. "In its present state," he wrote in 1968 of abdominal thermography, "thermography poses more problems than it answers" (Gros et al. 1968). Many of the specialists to whom Barnes and Gershon Cohen introduced thermography were unimpressed, and despite their efforts, and those of (other) manufacturers, the market emerging was a small one. Moreover, it was coming to be dominated by interests in breast cancer screening and, in effect, by radiologists.

The structure of relations corresponding to thermography's development phase was thus marked by the dominant, though not exclusive, position of radiologists and by low barriers to entry, which encouraged the tentative initiatives of a number of companies. These companies, looking to capitalize on their experience with infrared technology, were not part of the x-ray industry but were active principally in the defense area. As we now go on to look at the diffusion of thermography into practice, and at the assessments made of it, we will see that radiologists were not looking for a general imaging modality (as Howry had been in the ultrasound case). Their assessments were restricted to the utility of thermography in relation to breast cancer diagnosis and may be seen as connected to their professional interest in retaining control of an important area of their practice.

Diffusion and Accommodation: 1963 to the Late 1970s

As the phase corresponding to the integration of a new imaging technology into diagnostic practice, the diffusion phase is initiated by the first attempts publicly to specify its clinical uses. It is said to end with consensus being reached over at least some areas of the technique's clinical utility. So defined, thermography's diffusion and accommodation phase is very different from that of ultrasound. Recall that in the ultrasound case we saw the establishment of a set of interrelated but partially specialized markets. Accelerating growth took place within each of these markets, corresponding to identification of more and more uses within the various medical practices. We had a set of distinct dynamics, with none dominant. Neither radiologists nor the firms traditionally supplying them succeeded in gaining control of ultrasonic imaging, despite its medical and economic importance. Thermography shows a very different pattern, marked by continuing dispute and uncertainty. Here radiologists did succeed in appropriating the technology, though not in the obvious sense of unique control over its use. Rather theirs became the dominant *assessment* of the technology, and it was their assessment that largely shaped its future. The diffusion phase, as defined above, began *together with* commercialization. Indeed, initiated rapidly by enthusiastic "product champions," the attempt to establish clinical utility *preceded* any significant commercial activity.

Whereas thermography's development phase was short and unproblematic, its diffusion phase was neither. Recall that exploratory work had been inspired by interests in breast cancer diagnosis. Although in the development phase manufacturers sought other alternatives, it was radiologists who were coming to dominate the small market. And radiologists were never in any doubt that any possible utility of thermography would be in the area of breast cancer screening. So far as they were concerned, the research agenda was clear. By the time the first prototypes were available, Gershon Cohen in particular was insisting on its certain utility. He adapted his own diagnostic practice accordingly, creating a thermographic division within his radiology department in 1963, and thereby setting an example of the incorporation of thermography into radiological practice. In his hospital, all women reporting for a routine breast screening were thermographed as well as x-rayed from that time. The question was, of course, would the profession of radiology-at-large follow suit? Despite the uses emerging in other areas of medicine (e.g., rheumatology), it would prove difficult to reach agreement over the place of thermography in breast cancer diagnosis. Reflecting continuing controversy, early enthusiasm faded and the claims made for the utility of the technique gradually shifted. Something similar happened with regard to industrial interest. In the beginning, as we shall see, various firms possessing relevant expertise were interested in investigating its possible application to diagnostic imaging. By the end of the diffusion period they no longer were. Thermography in the 1970s showed a rather clear pattern of withdrawal and contraction. Radiologists gradually agreed that thermography was of little use in breast cancer diagnosis, as practiced by them, and the technique never penetrated beyond the margins of their practice. As far as both medicine and industry were concerned, the radiologists' highly negative assessment was to prove decisive. Though other problematizations, such as those of rheumatologists, based on quite different interpretations of and uses of thermographic evidence were to lead to other assessments, those of radiologists seemed to carry the most weight. And just as the radiological market for thermographic equipment stabilized at a very low level, industrial interest declined still further. As with radiologists, firms pulled out.

To begin with, as Gershon Cohen wrote in *Scientific American*, "The future of thermography as a diagnostic adjunct in all branches of medicine and surgery seems bright . . . thermography will add a significant new dimension to the diagnosis and prognosis of disease" (Gershon Cohen 1967).

The costs of building a few prototypes were not great, and there seemed little to lose. Apart from the companies that rapidly brought models to the market there were other companies with models at a prototype stage by 1966, including a few x-ray manufacturers, taking tentative steps: the French CSF company and Philips (Aarts 1977). Other companies that began to develop prototypes a little later included Fujitsu (followed quickly by other Japanese companies), Texas Instruments, and three British companies—Rank, EMI (for whom this first step into the field of medical electronics presaged the major commitment to CT scanning made soon afterward), and Barr & Stroud (makers of sophisticated optical instruments). Barr & Stroud completed their Videotherm in 1971 but never released it for commercial sale.

In illustration of the dynamic of a radiological/breast cancer research program in this period we can look at what was going on at the Albert Einstein Medical Center(AEMC), where women presenting for x-ray examination of the breasts continued to be routinely thermographed. From the roughly 3000 women per year examined, considerable data were accumulating. When Gershon Cohen retired in 1966, the thermographic work he had initiated continued, though the style in many ways changed.[7]

In 1969 a paper was published, based on analysis of data from all 2696 patients examined in 1967. Although within the program of work mapped out initially by the surgeons Lawson and Lloyd Williams, the research was clearly marked by its location within a radiological department committed to breast screening. To establish the place of thermography within radiological practice, it was no longer enough to *show* its utility for breast diagnosis. Radiologists wanted to establish utility more precisely, by comparing the accuracy of the technique with the accuracy of other available techniques. The 1969 paper posed the question which, for radiologists, must follow: How does thermography compare with mammography in terms of sensitivity and specificity? Statistical analysis of the data indeed suggested caution. Used alone, thermography could not distinguish benign from malignant disease "at our present level of knowledge" (Isard, Ostrum, and Shilo 1969). On the other hand, data did suggest that the rate of detection could be raised significantly above what either technique alone offered, by using them in combination. The paper used available clinical material statistically and carefully, to draw cautious and modest conclusions. It appeared that thermography could usefully be deployed *in conjunction* with mammography.[8]

The routine thermographic examination, the daily assemblage of clinical material, the generation of data for analysis, has not changed. What has changed is partly a matter of rhetoric, of style. But at a substantive level something more is changing. Like investigators elsewhere the AEMC group were being forced to confront serious and fundamental difficulties. Of these, the most serious arose from a growing recognition of the complexity of the "normal" thermographic pattern. If normal breasts were thermographically more complex than had been thought, how accurate was the symmetry criterion that was being used for diagnostic purposes? Could there be any confidence in the stability of the thermographic pattern, both over the reproductive cycle and in relation to longer term hormonal changes? For some time there had been those who had argued the importance of posing these sorts of questions before much sense could be made of large clinical series (see for example Branemark 1967). It was gradually becoming apparent that these difficulties could not be avoided.

The issue of the thermographic pattern of the normal breast had in fact been addressed a few years before by a group in one of Britain's major cancer hospitals, the Royal Marsden Hospital in London. Inspired by Lloyd Williams's work, Greening, a surgeon there, had begun work along similar lines. He selected a sample for examination consisting of 300 women with a lump in one breast plus, as control, 150 apparently normal volunteers (Harris, Greening, and Aichroth 1966). In a first report on their work Greening and his colleagues had written, "In assessing the value of thermometry it appears that a fundamental factor has been overlooked. Very little attention has been paid to the thermal patterns existing in the breasts of a normal, healthy female, and it is impossible to draw conclusions of abnormality without first having a background knowledge of normality." And there is the problem, for "after repeated examination of the 150 thermographs of normal women, we can find no system of classification."

Normal breasts, thermographically speaking, appeared to be very varied, with complex patterns of temperature distribution and lack of complete symmetry in thermal patterns. The criterion of symmetry as a means of diagnosing abnormality thus came to seem of questionable value. The problem here is akin to that faced by Kossoff (though actually a few years later) in his ultrasonic explorations of breast disease. When the normal control group was compared with the group having a lump in one breast, the immediate conclusions were ambiguous. A malignant carcinoma did produce a temperature elevation relative to the surrounding skin, but not necessarily corresponding in size and shape with the cancer. Benign tumors may or may not have an associated temperature elevation. Measurement of these thermal differences provided no clear means of distinguishing a malignant from a benign lump. To get further, Greening and his colleagues reasoned, you would have to step back from pathology and look more carefully at the *sources* of these thermal gradients. Certainly part of it is simply the distribution of veins, carrying warm blood, near the surface of the skin. An asymmetric distribution of these veins is likely to be the explanation of some of the asymmetry found in thermographs, and it would be necessary in some way or other to correct for this (figure 4.2).[9] The paper reaches the following conclusion: "It is an undoubted fact that some, if not all malignant tumors in the breast exist at a temperature higher than that of the surrounding tissues. Depending on various attenuating factors, this thermal gradient is reflected more or less at the skin surface where it can be accurately detected and measured by thermography. Superimposed on this thermal gradient are patterns of temperature resulting from superficial

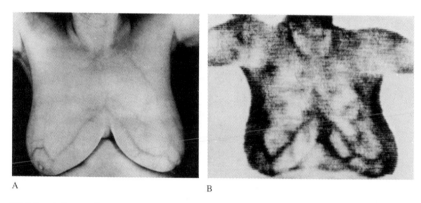

A   B

(A) Infrared photograph of normal subject with prominent subcutaneous veins.
(B) Thermograph of the same subject showing the effect of the venous pattern on the thermal pattern.

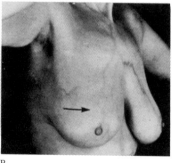

A   B

(A) Thermograph of patient showing thermal pattern similar to that asociated with prominent subcutaneous veins.
(B) Infrared photograph showing absence of veins in area indicated.

Figure 4.2
Thermal patterns of the female breast. From Draper and Jones 1969, 403. Reprinted by permission of The British Institute of Radiology.

veins. These patterns are generally asymmetrical and can augment or diminish comparative differentials of the temperature between symmetrically opposite areas of the breasts in association with a breast tumor." For this reason, pathology cannot be identified on the basis of comparing the two breasts with each other, since the assumption of normal symmetry does not hold.

This Marsden work had not had an immediate impact on radiologists. However, like investigators elsewhere, the AEMC radiologists were gradually becoming aware of the problems posed by the notion of the "normal" thermogram. It was becoming necessary to supplement the clinical data from routine examination with some more carefully designed experimental studies. Later, this would lead Isard and his colleagues to a reassessment of the role of thermography in breast examination. Other radiological groups responded differently to what, it was now coming to be agreed, was a major problem. After all, application of the technique had hinged on the use of this symmetry criterion in diagnosis. If complexity of heat distribution in *normal* breasts were such as to make this criterion problematic, radiologists were faced with a fundamental difficulty. Establishment both of the significance of thermographic images and, by the same token, of the utility of thermographic technique, was coming to require analysis that lay outside normal radiological problematizations and competence. It was not clear what future steps should be taken.

It was pointed out that while the criterion of symmetry "is the basic element of thermogram analysis, it has become increasingly obvious that it is valid only when equated with normal thermographic anatomy" (Dodd, Wallace, et al. 1969). There seemed to be a variety of factors that could make this baseline, the "thermographically normal breast," far too elusive for diagnostic comfort. One problem was the distribution of veins near the surface of the breast. These, "as well as anatomic dissimilarities, may produce a spurious 'hot spot' which can be analyzed properly only when adequate spatial resolution is obtained. No detailed study of normal thermal patterns is yet available."

This formulation of the problem suggests two possible research strategies. One, implied directly, is the need to try to understand this normal baseline for diagnosis. As a research problem, this was at the fringe of radiologists' interest and competence. Alternatively, if the problem was essentially one of the adequacy of spatial resolution, then it could be seen as in part one of the *design of the apparatus*. Perhaps the designers of the Pyroscan (the instrument this group was using) had gone too far in sacrificing image quality to scanning speed. In the University of Texas M. D. Anderson Hospital, G. D. Dodd and other radiologists were beginning to "reproblematize" their thermographic work: to set it within a different sociocognitive context. At this time Dodd was in contact with the Dallas-based Texas Instrument Company.[10] Within this "transepistemic arena" the instrumentation as such was made the subject for consideration, a strategy that fitted well with Texas Instrument's interest in new markets.

By late 1968 tests of a new experimental thermograph developed by Texas Instruments were beginning at M. D. Anderson. Although it had the same liquid-nitrogen-cooled detector system as the Pyroscan, this instrument was designed to provide for higher spatial resolution, with the scan time being allowed to rise correspondingly. Improved spatial resolution was going to be the selling point, and Dodd and his Texas Instruments collaborators stressed the significance of this in reporting on the new device. Writing in *Cancer* in 1969, they pointed out that "One of the most difficult aspects of interpreting thermal scans has been the lack of clear anatomical detail. Obviously, the ready identification of normal structures permits ready recognition of abnormalities" (Dodd, Zermeno, et al. 1969).

An interesting shift was thus attempted. The awkward concept of "thermographic normality" was abandoned, bringing the focus of the technology back to (radiologically) familiar territory with "improved spatial resolution" and *normal structures*. Structures, of course, are the very essence of radiological practice, and Lawson had initially expressed the interest of thermography precisely in terms of its representation of process rather than structures. This shift is clearly to be understood not only as a way of arguing for the benefits of the new instrument but also as a step in the assimilation of the technique into radiology.

Moving into the 1970s, the sentiments surrounding thermography—in essence the significance of thermography for the medical profession—evolved further. As those committed to thermography fought a rearguard battle against an emergent consensus within radiology that the technique had little value, the argument continued to shift. Increasingly, however, debate took place in a climate marked by frustration and even bitterness.

Reviewing the status of breast cancer detection in 1977, Dodd suggested that at the beginning of the 1970s thermography seemed to have a clear though modest screening role (its position "appeared reasonably well established"), particularly given the fears of overuse of x-ray. There were hopes that specificity and sensitivity could be improved (Dodd 1977). In line with this assessment, the British Department of Health and Social Security felt it worth approaching the Ministry of Defense Research Establishment at Aldermaston and asking them to design and build a prototype device specifically for the detection of breast cancer. (This prototype was completed in 1974/75.) Subsequently, an assessment of the technique was attached to a twenty-seven-center demonstration study focused on mammography and under the auspices of the American Cancer Society and the National Cancer Institute. This should have confirmed its strengths and weaknesses, thereby placing use of the modality "on a firm footing." But it was not to be: the evaluation was very much more negative than expected. Instead of forging a consensus on a modest role for thermography in screening, the evaluation led to the National Cancer Institute ruling the technique out as almost useless. A British National Health Service evaluation also carried out in the 1970s reached somewhat similar conclusions. Criticism of the adequacy of these particular studies has become an article of faith among thermographers today.[11] One report on the British study suggests that "to say the least it was ill conceived. . . . Little or no thought had been given, through inexperience to the necessity to stabilize the patients temperature and to take account of the female monthly cycle . . . deodorant powders and other cosmetics caused interference."[12]

In the face of growing skepticism and withdrawal, the only strategies open to groups that had committed themselves to breast thermography were defensive. What strategies were possible, given the constraints imposed by radiological practices and by demarcations of professional competence? One defensive strategy, adopted by the AEMC group, was to rethink the place of thermographic imaging in diagnostics. Another, adopted by their neighbors in Philadelphia at Temple University, was to rethink the diagnostic criterion used. Thus when the AEMC group came to report on the 10,000 women examined in the four-year period from 1967 to 1970, the claims made were different from before. What exactly is thermography going to mean for breast cancer diagnosis? "To avoid unfortunate misinterpretation, it must be clearly stated that thermography cannot and does not *diagnose* cancer. Its use is analogous to temperature recordings in the course of studying any disease process" (Isard et al. 1972).

This is very far indeed from Gershon Cohen's claims five or six years previously. It had seemed that, locally at least, the meaning, the clinical significance, of thermography in breast examination had been securely established. But Isard asserted that, "The thermogram at this stage of development of the discipline of thermography can no more by itself differentiate a benign from a malignant condition than can the temperature recording by the oral thermometer differentiate pneumonitis from a necrotizing neoplasm."

The most credible claim that could now be made for the value of thermography was as a useful *preliminary screening modality*, to characterize a group meriting further investigation by other techniques. Its innocuous character, the fact that it makes no use of radiation—scarcely referred to before, and an awkward selling point for radiologists—now became a crucial advantage. At the same time a number of frequently noted difficulties had to be acknowledged, including the high false negative rate (low specificity) of the technique. More than a third of the 36 cancers found by biopsy among the asymptomatic group had produced negative thermograms. Hardly to be ignored, this was better confronted: "We do not yet understand whether the preservation of a normal pattern is a reflexion of the size, location, or biologic behavior of a lesion, whether it is related to the nature of the underlying breast matrix, or whether interpretation of the pattern as normal is in error, because some subtle changes defy recognition at this stage of our experience" (Isard et al. 1972).

At Temple University's Diagnostic Radiology Research Laboratory, Ziskin and his colleagues were engaged in an ambitious attempt at computer analysis of breast thermograms. The objective was a formal diagnostic criterion, a defensible means of recognizing pathology, since the simple criterion of departure from lateral symmetry had been discredited (Ziskin et al. 1975).

The procedure was first to digitalize the thermogram into "192 × 256 picture elements called pixels, each of which is assigned one of 512 gray level values" depending upon its brightness. Features of interest were then defined, which "represent our crudest concepts of what is important *about* a thermogram visually: "hottest region on the left," or "average temperature of right breast," or "number of hot regions on the left." Each of these features was given a P-number, and the P-numbers were combined to form complex features "suitable for input to the diagnostic program....These parameters are passed to a statistical decision program. This program has been trained on our thermograms to provide a set of classification coefficients . . . chosen so that maximal separation between the normal and abnormal classes could be attained. The output . . . is a diagnosis and a probability of error in diagnosis for the specific trial."

What this work amounted to was an attempt to set diagnosis free from the difficulties that beset it by redefining pathology. Pathology would no longer be defined in terms of (falsifiable) assumptions as to thermal correlates of malignancy but instead in terms of "that [read "whichever"] coefficient has in practice proved best able to discriminate normal from pathological." Such a purely phenomenological diagnostic criterion, without physiological or similar referents, can in principle be endlessly improved. Because it says nothing at all about disease, does not rest upon any assumptions regarding the nature or etiology of the disease, it is not to be undermined by progress in pathology.

For nonradiologists concerned with breast cancer the situation was rather different. At the Royal Marsden Hospital in London, just as at the AEMC, women attending the Well Women clinic for screening were being routinely thermographed. But the research published was of a quite different sort, largely focusing neither on the accumulating clinical data nor indeed on the signs of pathology. Questions outside the interests and research competence of radiologists were precisely of interest for this group, composed after 1968 of medical physicists and surgeons. What occurred were careful *experiments* designed to explore the problem of thermographic "normality." In 1969 Draper and Jones, two medical physicists working at the hospital, had been able to provide a classification of the variety of "normal" thermographic patterns, using infrared photography to correct for vein patterns. Analyzing the thermographs of 442 clinically normal women, and taking account of such possible complicating factors as age, menstrual cycle, and use of oral contraceptives, four essentially different basic patterns were identified (see figure 4.3). Normality may be very complicated, Draper and Jones were essentially saying , " but now we can

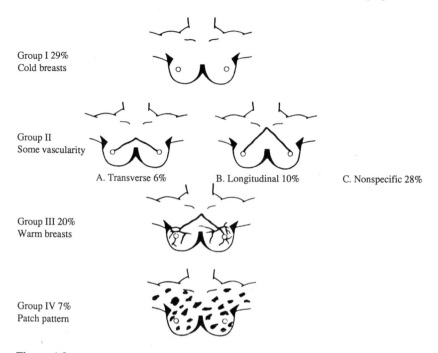

Group I 29%
Cold breasts

Group II
Some vascularity

A. Transverse 6%   B. Longitudinal 10%   C. Nonspecific 28%

Group III 20%
Warm breasts

Group IV 7%
Patch pattern

Figure 4.3
Classification of normal thermal patterns. Draper and Jones 1969, 404.
Reprinted by permission of The British Institute of Radiology.

grasp it," now we might be some way toward a more meaningful diagnostic criterion. The thrust of their 1969 paper had not been that the technique should be abandoned; were it to be, there would be little future in their own research. Just like the AEMC people at about the same time, the Marsden group had suggested that thermography probably had a principal use as a screening technique, by virtue of its "simplicity and speed" and the "lack of any comparable screening method." Work published at the beginning of the 1970s by the Marsden group continued to deal with the themes of characterization of the normal thermogram (Jones and Draper 1970) and of the processes by which thermal patterns arise (Draper and Boag 1971).

In 1975 a report was published on the total five-year series of 12,000 women, of whom 8000 were healthy and symptomless. From this large series, the group having cancer at a known stage of development could be selected out: 342 women. From these data it appeared that early stage tumors caused detectable thermographic changes in the diseased breast less frequently than did advanced tumors. "This difference accounts for some of the variations between the results of workers who have not considered the effect of the stage of disease on the thermographic assessment" (Jones, et al. 1975). Jones and his colleagues then reopened discussion of the significance of thermographic evaluation in two distinct ways. The first argument involved a redefinition of the diagnostic criterion, though in a quite different fashion from the phenomenological approach of the Temple radiologists. It was clear by then that simple departure from symmetry was no basis for diagnosing pathology. What remained was the possibility of serial assessment, of change over time, made possible by the hazardless nature of the technique. Second, a reevaluation of the significance of thermographic findings was attempted by shifting the frame of reference from the presence of a tumor (as confirmed by mammography, or by biopsy) to *survival* "patients with Stage II or Stage III disease who had normal thermograms tend to have better survival rates than those with abnormal thermograms."

Logically speaking, since it proposed a radically different way of interpreting thermographic findings, this paper should have set the scene for renewed interest. Yet the Marsden workers published nothing further on breast thermography; interest in the subject simply fizzled out.[13] Such a notion is of course far from the conventional picture of scientific experimentation involving, in its more idealized forms, a clearly formulated question and answer. Clinical diagnostics allows for a quite distinctive notion of "how experiments end." It is not that thermography was abandoned. Women continued to be routinely thermographed (as they still are). Data continued to be collated in the computer, the consequence of a certain logic in the screening procedures. Routine thermography continues, according to Dr. Jones, because clinicians seem to find it useful to have a baseline thermal image, something that can be checked without risk on each occasion. Further collection of thermographic data made sense within the local context because equivalent data had previously been collected for these women. The data provided a means of keeping an eye on individual patients more regularly than would be acceptable with x-ray. But the significance of these results was not to be generalized beyond the local context. By the late 1970s their relevance for breast cancer diagnosis and screening could no longer be argued in print in a way which radiologists would find convincing, and the practice had to remain localized. So no one found time to write up the accumulating data.

There is however more to the career of thermography in this period than this. Though its utilization in the diagnosis of breast cancer was the principal interest for radiologists, other applications had been promising enough to be further investigated. Rheumatologists had turned to it with concerns that would make rather different demands on the technology. Whereas the breast researchers were seeking a means of early *diagnosis*, of deciding whether pathology was present or not, for the rheumatologists the association between pathology and heat was unproblematic, hallowed by centuries of medical history. They wanted a means of accurately *monitoring the progress* of the disease. We shall see that, in contrast to the breast thermography work, the 1970s were a period in which the rheumatologists were able to build on technical and social accomplishment, on a positive consensus. There is a striking contrast to the ultrasound situation in which partially independent markets emerged and radiologists failed to gain control of the technology. In the case of thermography, radiological failure has dominated success in rheumatology.

Leaders in this direction were the group at the Royal National Hospital for Rheumatic Diseases in Bath. Early studies focused on establishing characteristic thermal patterns of normal and arthritic (i.e., inflamed) joints, in particular the knee (Cosh 1966, Cosh and Ring 1967). Uses that the technique might have in rheumatology were clear enough from the start. "Thermography is of use qualitatively in illustrating patterns of heat and cold in the skin of normal and abnormal subjects. In rheumatoid arthritis it is informative in showing heat patterns overlying inflamed joints, and it may indicate which part of a complex joint, such as the knee or wrist, is mainly affected. It illustrates well the relatively rapid changes which follow articular steroid injection. When adequately standardized . . . it will be of value in recording the long term progress of inflammation in affected joints" (Cosh 1966).

Although there were certainly difficulties at this early stage, the technique seemed very promising. Presenting their first results to an international thermography symposium held in 1966 in Strasbourg, the Bath group were therefore surprised to discover that no one else in Europe appeared to have seen a use for the technique in rheumatology.

Because at that time the Pyroscan did not include any means of standardization, absolute temperatures could not be measured—only temperature differences. This was a limitation. It meant that in quantitative work, which was assuming greater and greater importance in rheumatology, other techniques had to be used in conjunction with the Pyroscan. However, the combination of thermograph and radiometer proved "superb." Dr. Ring explains, "The Pyroscan showed us that there were patches of heat in certain parts of the body, but we had no means of quantifying that. The radiometer did that for us. So by combining this crude imaging technique with the radiometer we got a *feel* for quantitation. And that really laid our foundation for the quantitative thermography work which has occupied most of our time since" (interview with Francis Ring).

Quantitative studies were coming to be used in particular as a means of monitoring steroid injections and other treatments, with first results published in the *British Medical Journal* in late 1968 (Ring and Cosh 1968). It is important to realize that the technique was being used in a quite different way from the breast cancer studies. Quantification was important because the rheumatologists wanted to use thermography not to diagnose arthritis, but to monitor the progress of the disease over time. A second area of interest was also developing. Contacts had been established with the pharmacologists from nearby Bath University, and there was a common interest in the relationship between thermal and biochemical changes in the inflamed joint (Collins and Cosh 1970). Quantitative thermographic scanning would facilitate understanding of the physiological and biochemical processes taking place, so that the two sets of interests could usefully be brought together.

Their interests in quantification had rather clear technical implications, and the Bath group tried out other instruments in addition to the Pyroscan, including instruments produced by AGA and Bofors. Such interests meant, in effect, different priorities and different trade-offs in the choice of instrument from those the cancer researchers were making. For example, speed of imaging was less important, thermal resolution more so.

By 1970 the significance of accurate measurement of temperature elevation in the investigation of arthritis was known and accepted. The established relationship between thermal and biochemical changes gave added significance to precise thermal measurement. All indications pointed to a clear clinical use for thermography in the assessment of rheumatoid arthritis and of its treatments. Though questions remained, not least relating to performance of equipment and the approach to quantification, these assumed the value of the technique as such. Research was in full swing, with clear achievements and clear objectives for future attention.

Work gradually came to focus on the development of a means of assessing quantitatively the progress of rheumatic disease and, later, of assessing the effect of therapeutic procedures and drugs. In 1974, working with Bath University pharmacologists, the rheumatology group announced their " thermographic index" in the *Annals of Rheumatic Diseases* (Collins, et al. 1974; Ring et al. 1974). What these papers attempt was, first, the presentation of a quantitative "index" of rheumatic disease and, second, its validation. Thermographic data from normal knee joints and from the knee joints of patients known to be suffering from rheumatoid arthritis were first displayed isothermally.[14] It was then possible to measure the areas within each isotherm; the point was to see how much of the skin surface was at various temperatures. Validation of the index, the attempt to convince professional colleagues of its utility, was achieved by reference to a common treatment. The writers showed how injection of steroid drugs in arthritic knees, an accepted therapy, brought about a reduction in the thermographic index. A later step was to show that the index could be used to measure the action of a drug not only under controlled research conditions but also in routine practice. The index had been designed for the evaluation of a new treatment under research conditions as well as for assessing the progress of the individual patient. These complementary though overlapping functions involved the Bath group in different sets of problems, both technical and social. On both fronts they were successful.

Through much of the later 1970s, published work was concerned with the evaluation of new procedures and drugs using the thermographic index (together with clinical assessments and laboratory measurements). Important results, in Dr. Ring's view, turned out to be conclusions such as "the smaller the joint the faster it responded, both to being taken off the drug and to being put on it," and that "aspirin and aspirin derivative drugs in a big enough dose could lower the thermographic index." These conclusions attracted the attention of the pharmaceutical industry, with which the group subsequently developed very fruitful links.

Validation of the thermographic index was so successful that in 1977 AGA (which had come to occupy a dominant position in the small thermographic market that existed) produced a device known as the "AGA integrator," an accessory to their Thermovision, which was supposed to provide "the basic facilities for calculating the thermographic index at a fraction of the cost of a computer" (Ring 1977).

The careers of thermography within the two practices of (radiological) diagnosis of breast cancer on the one hand and rheumatology on the other were very different, entailing quite different processes of evaluation and adjustment. That the former, with its gradual loss of optimism and final dismissal, was the more important can be seen by looking at changing industrial involvement in this period.

The commercialization of thermography from about 1970 was marked by growing industrial hesitancy and a common reluctance to make commitments beyond the construction of a few prototypes. Both Rank and EMI built three or four prototypes but were unwilling to solve the (different) technical problems that inevitably arose. The EMI instrument seems to have had very poor resolution. Ring recalled that "it just died a natural death." On the other hand, Rank "were actually ahead of their time, because they, working with military contracts, had got hold of a multielement detector which was capable of working at conventional video speed . . . and this we found to be very exciting. And it was with that particular camera that we, working on an assessment project for the DHSS here in Bath, Lloyd Williams and I took interest in possible *dynamic* studies."[15]

Unfortunately the instrument required very careful and expert use, and there were many problems of "shifting and drifting . . . the picture got very broken up." But in the right hands, "the Rank equipment was running years after the other prototypes. And Colin Jones was using one, and we were using one, for many years . . . you could get a real time image, of sorts."

It was not only that firms like Rank and EMI chose not to go beyond the prototype stage. Both Barnes Engineering and Smiths, makers of the Pyroscan, also decided to drop out. "It was very disappointing," comments Francis Ring, "having made such a sensation in New York in the early sixties with this 'high-speed scanner' as it was then termed, Smiths had got on the drawing board a CRT display system which would be far more flexible and adaptable, smaller, more efficient. But the financial implications were too much for the company. And unfortunately they were not able to obtain more money from the DHSS." (Interview)

Smiths decided not to press ahead with their improved thermograph because, by the beginning of the seventies, the investment had become too great a commercial risk. Moreover, as Ring explains, the DHSS was unwilling to share this risk.[16] Unable in this way to reduce the uncertainties of further investment, Smiths terminated their medical infrared activities, selling their stock of ten or so remaining Pyroscans to the DHSS, which gave them out on free loan to those who applied. In Ring's view this is more than just a sad ending to the story of the Pyroscan. It also had significant implications for the general view of thermography that more and more came to dominate in British medical circles.

That was going on while the improved AGA and Bofors cameras were coming in from Sweden, but few people could afford to buy them. We were then talking about £5000–7000 for an AGA or a Bofors. And you could have an old Pyroscan from the DHSS on free loan. So people were very frustrated. . . . There was a company . . . in London which started to import the Bofors camera, and AGA set up their own company here. They were . . . loaning their equipment to people in order that papers would be published and the superior pictures would appear in them . . . Both of them, in competition one against the other, against the background of growing dissatisfaction with the old early English equipment which was slow and . . . so on. And this, in my opinion contributed to the beginnings of the bad press of thermography in the UK. (Ring interview)

The Feedback Phase: From the mid 1970s to the mid 1980s

Diagnostic ultrasound was a convincing example of how success leads physicians and industrial scientists alike to search for ways of improving both an imaging technology and the range and scope of its applications. At each stage positive signals provided encouragement, hinting at further possibilities yet to be explored. Its recent history is marked by a very rapidly growing market, the continuous search for new uses, major investment in R&D, and a high level of publishing. By contrast the history of thermographic imaging is a negative example of this process. Radiologists had assessed the utility of thermography for their practice not as a general imaging modality but in terms of a very specific set of questions relating to breast cancer diagnosis and screening. This approach to assessment derived from early suggestions that the technique might be of unique value there, potentially posing a threat to current practices in breast screening. By the mid 1970s radiologists had agreed that the technique neither threatened nor usefully complemented their existing practice. Indeed, their evaluations of thermography seemed to hint only at the limited payoff to be expected from further investment of time and effort. Subtle attempts were made to save the technique through a variety of strategies which could be developed within the bounds of agreed radiological competence and conventions. We saw that these included recourse to a phenomenological notion of diagnosis in one center and a shift in the frame of reference from the presence of a tumor to survival rates in another. Industry, however, responded not to these subtle strategies but to the overall consensus. Barr & Stroud's experience shows how this worked.[17]

When the Aldermaston research establishment produced their thermographic device for breast diagnosis in 1974/75, the health ministry (knowing of their previous interest) asked Barr & Stroud to take on its final engineering and marketing. The firm agreed but, having no knowledge of thermal scanners (their Videotherm was never sold), the firm produced the first batch of IR11 devices strictly according to the Aldermaston drawings. A small number of instruments were sold, but the design was beset by difficulties. "The equipment was very large in size and one of the features the Health Service had requested was a fixed focus which was provided in the form of a Barnes-Wallis rangefinding technique where the camera was pushed toward the patient until two spots of light were in coincidence. This fixed focus produced severe limitations on the camera being used for other medical applications and the weight and size of the camera made it very unwieldy and difficult to operate." (D.C. Lorimer, personal communication)

Design specifications which had been derived from the requirements of breast cancer screening thus made alternative uses of the instrument problematic. This, clearly, is one reason for which the radiologists' assessment became decisive for industry. Despite the demonstrable utility in rheumatology, the late 1970s were marked by a contraction of what little support for thermography had been built up.

Nowadays industry is wholly uninterested or uninvolved or, in some cases, simply unwilling to countenance the last step into the market. A report on the diagnostic imaging industry in the United

States writes of thermography: "Very little, if any, research is being done by industry in this field. Honeywell is collaborating in a government-sponsored program in developing a system involving computerized infrared pattern recognition. However, it is believed that Honeywell will not commercialize their computerized system" (Hamilton 1982, 83).

In the late seventies Barr & Stroud developed an infrared camera, initially for military purposes, far in advance of anything available on the medical market. "Very considerable interest was shown in the IR18" and the company became closely involved with a small California-based company called Micro Thermology, dedicated to providing advanced thermographic equipment to breast screening clinics. When, there too, early promise fizzled out Barr & Stroud decided not to go ahead with commercialization of the IR18 for medical purposes, seeing no market in the near future. Such a situation, in which the commercial advantage attaching to unique technical expertise seems not to merit exploitation, would be unimaginable in the ultrasound case. Feedback processes have effectively ended, since industry is not interested.

For those involved, still committed to the possibilities of thermography, the endless question is, why and how did it happen? "The question therefore must be, why has medical thermography not been accepted virtually anywhere in the world after the enormous amount of work that has been put into it by the medics, science and industry. It would appear to me that it is believed by the majority of physicians skilled in thermography that the reasons for the marginal results obtained in this program are poor training of screening center personnel in the use of the technique, poor attitude, poor protocols, poor quality control, incompletely defined interpretation criteria and most importantly a general misunderstanding of the proper role of thermography in breast cancer detection and other disease" (D. C. Lorimer, personal communication).

A sense that the technique was inappropriately, or prematurely evaluated, is a belief that unites the small band of thermographers. Some see hope in technological advance: "The nonspecificity and lack of sensitivity of current instrumentation should be the impetus for further advances in technology. The concept of a total body thermogram as a routine survey for areas of local increased heat production or deficiency is not without merit; and the thermogram can find its place, as did the thermometer, in the study of human diseases" (Freundlich 1980).

Others hope that consensus will ultimately emerge as to the *prognostic* value of the technique in the breast cancer field. Recall how the 1975 Marsden paper had suggested that within a given tumor class, patients with normal thermograms seemed to have better survival chances than those with abnormal thermograms. In France, Gautherie and his colleagues in Strasbourg have done more work on this point (Gautherie et al. 1983) and evidence is accumulating. In July 1987 Dr. Isard was about to publish American data suggesting just such a prognostic value, though he was unsure as to whether they would be accepted.[18] However, Dr. Isard's strategy would move deliberately outside the bounds of radiological convention: deploying—rather than being constrained by—demarcations of professional competence. Isard had decided to present his results to surgeons, not his fellow radiologists. His strategy was to persuade the surgeons to ask for a thermogram, so that radiologists would have no choice but to provide one.

Apart from a general view as to the organization of the medical world, Dr. Isard's strategy is also based on features specific to his reading of the history of breast thermography. In France and Italy (together with Japan and the Soviet Union) thermography has never fallen into quite the disrepute charted here. According to Dr. Isard, this has to do with preferences in the treatment of breast cancer. In France for example, there had always been much less recourse to mastectomy (an operation widely carried out by surgeons in the United States) and much more use of radiation therapy. In monitoring the decline in radiation-induced heat, thermography had an obvious use that it lacked in the American therapeutic regime. However, slowly American practice was changing, and Dr. Isard seemed to judge the time opportune.

But so long as manufacturers see no commercial future for the technique, this optimistic belief in technological advance is sadly misplaced. As D. C. Lorimer from Barr & Stroud writes, "no commercial company can remain in business by financing the development and the provision of equipment as has been done over the past 15 years with no prospects of recovering these costs through eventual sales."

Thermographic imaging was conceived in response to Lawson's finding of a "hot spot" over a breast tumor, which led him to search for a sophisticated means of skin temperature measurement. The rhetoric associated with the origins of thermography stressed its fundamental difference from the roentgenographic representation of anatomy—thermography portrays function, process. It was undoubtedly the widely recognized importance attaching to early cancer detection and to breast screening in the 1950s that provided the essential stimulus to the transfer of expertise. Firms with knowledge of infrared technology were almost all in the defense field. We can imagine that breast imaging, then becoming a major commitment of health systems (with modern mammographic equipment recently developed) appeared to be a distinctly attractive possibility for moving into medical markets. Entry barriers were low; it cost little effort to make a thermal camera specifically for diagnostic purposes. Though other uses than breast cancer diagnosis were sought, and some were found, this earliest incentive always dominated industrial interest and shaped design decisions. Breast screening had become an integral and important part of radiological practice, and it was radiologists who soon dominated the assessment. Not the more general diagnostic possibilities of thermography, but its place within the context of existing radiological approaches to breast examination formed the basis for their assessment. The early sense that thermography was not to be judged relative to (other) modes of representing structure was swept away. When the thermal patterns of normal breasts were found to be highly complex, so that the diagnostic criterion of departure from symmetry ceased to be tenable, radiologists largely lost interest. Search for the meanings of these patterns, for the true thermal correlates of pathology, lay outside their interests and research tradition.[19] The technique offered neither threat nor promise. It is this sense of failure, dominating radiologists' perception of thermographic imaging, that has shaped the response of industry. Even though the few firms that took tentative steps in the thermographic imaging area were not established suppliers of imaging equipment, nevertheless their perception of the market has been dominated by the opinion of radiologists. The market that has stabilized, principally in the unfashionable medical specialties such as rheumatology and burns therapy (with a minor and highly localized practice in breast imaging), sustains only minimal industrial interest. Attempts to save the technique by research strategies carefully crafted within the constraints of radiological convention and agreed demarcations of competence have all failed. Dr. Isard's current strategy, to outmaneuver the radiological profession by ignoring conventions and appealing directly to the treating physicians and surgeons, is a novel utilization of the social structure of medicine. Whether it will succeed, whether thermographic imaging may yet find a secure place inside or outside radiological practice, as ultrasound did, is hard to say. It does not seem likely.

5

Into the Digital Age: The CT Scanner

Introduction

Computed Axial Tomography (CAT) scanning—now more usually known as Computed Tomography (CT) scanning—was conceived under very different circumstances from those surrounding the techniques that preceded it. Its industrial origins, the planning that went into its conception, and the rapid buildup of a ferociously competitive market explain, in part, why its career was to prove so different from those of ultrasound and thermography. But only in part.

The career of CT scanning also differs from earlier imaging technologies because it was, as it were, born into the very different world of modern electronics and computers. Both diagnostic ultrasound and thermography had largely taken shape by the early 1960s. Although the first phase of the electronics revolution, marked by the introduction of transistors, had begun in the mid fifties, military applications for long remained the principal market (Braun and MacDonald 1982, ch. 6). Technical limitations and cost initially held back the range of applications, but the pace of development was rapid. A rapidly emerging use for transistors was in the new computer industry, for by replacing tubes with these new semiconductor-based devices both size and power consumption of computers could be dramatically reduced. In 1959 Digital Equipment Corp. marketed a relatively small and inexpensive "minicomputer" "dedicated" for use with all sorts of laboratory instruments (Laitanen and Ewing 1977, 335). But this was only the beginning.

The second and perhaps more dramatic phase of the microelectronics revolution came with the introduction of integrated circuits. A single tiny wafer—first of germanium, later of silicon—

could substitute for more and more transistors and other components. The number that can be put onto a single chip is said to have doubled every year since 1960 (Braun and Macdonald 1982, 103). We saw how this rapid improvement in electronic components was a reason for hesitation in bringing an ultrasonic device to the market. In the late 1960s Nuclear Enterprises was understandably worried that their product would rapidly become obsolete. When in November 1971 Intel announced the first "microprogrammable computer on a chip," what we now call a microprocessor, the way was clear for the computer revolution and widespread cheap computing power.

In recent years the computer storage and processing of diagnostic data has become common in medicine, especially for research purposes. But though computers are now used in the processing and analysis of both ultrasonic and thermographic images (recall Ziskin's phenomenological analysis of breast thermograms in the mid 1970s), they are not fundamental to the techniques of either ultrasound or thermography. By contrast CT scanning, as its name suggests, is wholly inconceivable without modern computer technology. It has been able continually to profit from advances in that technology.

The theoretical problem that led to the computer being brought to bear on diagnostic images was not a purely medical one. It had arisen in similar form in scientific disciplines like radioastronomy and x-ray crystallography, which also involve mapping a complex structure in two or three dimensions (Edge and Mulkay 1976). In 1917 an Austrian mathematician, Johann Radon, had published a paper in which he proved that a two-dimensional or three-dimensional object can be reconstructed uniquely from the infinite set of all its projections. Application of this work, which was to take place in numerous fields—including radio astronomy, crystallography, and microscopy as well as radiology, could only take place with the development of the modern computer, since enormous numbers of calculations were required for each reconstruction (Gordon, Herman, and Johnson 1975).[1] It is not difficult to see this. Following the example given by Gordon et al., suppose we want to reconstruct the structure of the brain as a cube with sides measuring 20 cm. For diagnostic purposes we may need to be able to resolve to 2 mm accuracy, so that the brain would be reconstructed as an array of small cubes, each with 2 mm sides. If the computer took only one twentieth of a second to make the computation for each small cube, the total brain ($100 \times 100 \times 100$ small cubes) would require more than half a day of computing time. In fact, as we shall see, application of this principle was achieved through the development of algorithms that vastly reduce this computing problem, made possible by the rapid development of computing in the 1960s and 1970s.

Problems of this general kind had arisen long before in radiology, though they had been tackled with wholly other means. Early in the history of the field certain inadequacies in methods of localizing pathologies in the body had been noted. Where an organ to be visualized was obscured by a bony structure, or where it became important to try to assess how deep within the body a pathology was to be found (thus localization in three dimensions), existing technique was soon found to be inadequate. In the 1920s and early 1930s a number of radiologists (including the Frenchman A. E. M. Bocage and the professor of radiology in Amsterdam, B. G. Ziedses des Plantes) developed the technique of tomography. *Tomography* entails moving the source of x-rays relative to the body in one direction while at the same time moving the film in a parallel plane in the other direction. In this way the projected image from only one plane within the body remains stationary relative to the sensitive film. The consequence is that the image from this plane remains sharply focused, whereas the images from all other planes are (deliberately) blurred. In the early thirties the German radiologist Gustav Grossman developed the first tomographic instrument to be put onto the market (Süsskind 1981, note 7). This tomographic principle became the traditional means by which radiologists obtained three-dimensional information.

The Exploratory Period of CT Scanning: 1961–1973

It is generally agreed that the first to consider new methods of handling x-ray information was University of California neurologist W. H. Oldendorf. In a paper published in 1961 Oldendorf pointed out how profound dissatisfaction with existing methods of brain imaging—angiography and ventriculography—had led him to consider other alternatives.[2] "These tests were introduced into clinical medicine between 30 and 40 years ago, and neither has changed basically since then. Each time I perform one of these primitive procedures, I wonder why no more pressing need is felt by the clinical neurological world to seek some technique that would yield direct information about brain structure without traumatizing it. It was this firm conviction that prompted the development of a system which is theoretically capable of producing a cross-sectional display of radiodensity discontinuities such as the head. At the time of writing no biological system has been studied by this method" (Oldendorf 1961).

CT scanning's exploratory period was in a sense initiated by a clear "need pull." But although Oldendorf went on to construct a working model, and in 1963 to take out a patent, he was unable to interest either radiologists or industry in his idea and nothing came of it. The idea of measuring transmission profiles and then using these to calculate linear absorption coefficients was subsequently taken up by a Tufts University physicist, A. M. Cormack. Using a weak gamma ray source, Cormack carried out an experiment on a wood and aluminum phantom and found that his method yielded gamma ray absorption values close to known values for these materials. Noting the possible application in radiology (for determining variable x-ray absorption coefficients in two dimensions), Cormack published his findings in the *Journal of Applied Physics* in 1963. They went wholly unnoticed by the medical community (Cormack 1963). It was Godfrey Hounsfield, a British electronics engineer working in industry, who a few years later was to succeed in bringing all the interests together and is generally credited with invention of the CT scanner.

The problem they all shared, and which also confronted radioastronomers, crystallographers, and electron microscopists, was that of reconstructing the two- (or in principle three-) dimensional image of an object from a set of its projections.

Godfrey Hounsfield had been a radio mechanic in the war.[3] The capabilities he showed working with radar were such that after the war he received a government grant for further training. Graduating as an electrical engineer in 1951, he joined Electric & Musical Industries Ltd. (EMI), in Middlesex.

EMI traditionally made records and home entertainment equipment, having pioneered electric (in place of mechanical) recording in the 1920s and television in the 1930s. A research laboratory had been established in 1928/30. After the war, though the entertainment business was to remain the firm's stock-in-trade, there was a decision to branch out into both civilian and defense-related electronics. Hounsfield was set to work to develop a computer for business applications. By introducing transistors wherever possible,

EMI hoped to produce a better machine than the only other business computer (constructed with tubes) being made in Britain.[4] Great hopes were placed on the EMIDEC 1100, and with an initial order for two dozen machines, dramatic entry into the emerging computer market seemed assured. However, rapid growth in the industry, and a growing sense of what would be needed to keep up in terms of investment, determination, and sales and marketing staff, made it apparent that the firm would not be able to stay at the forefront in computing as well as in its traditional business. In that area too there were great opportunities, rapid technical progress, and an enormously buoyant market. EMI management eventually decided to withdraw from the computer market and the EMIDEC 1100 designs were sold to the computer manufacturer ICL.

Hounsfield moved to the company's central research laboratory. His first assignment, which involved improving accessing of computer memories—it was hoped that patentable and licensable discoveries would emerge, came to naught because computer manufacturers were not interested. At that point Hounsfield was asked to consider what his future line of work might be. He made a number of suggestions and it was eventually agreed that he would work on pattern recognition.

Problems of pattern recognition—the automatic identification of alphanumeric characters and other patterns (voices, representations of structures obtained through research with electron microscopes, etc.)—arose in many fields. How could you best scan characters and transfer information from one form to another? Süsskind quotes from his interview with Hounsfield: "If you scan characters with straight lines, many of them fit—E's for example. If you can find a variety of shapes with which to scan, then you do much better than scanning with a spot in a raster form [as on a TV screen]; this brings down the amount of work to do. That was something I was working on at the time" (quoted Süsskind 1981, 47).

One possibility Hounsfield was considering, perhaps an association from his work on computer memories, was this. How would it be if, instead of presenting the information derived from a pattern instantaneously, it were first stored in a computer's memory and then processed to cut out redundant information? Thinking about this, "during a long country ramble . . . the basic idea of CT scanning started to form in his mind." At a fairly early stage it seems to have occurred to him that his idea could be applied to roentgenography. In principle, as Radon had shown, if you made measurements through an object from many angles, you ought to be able to collect enough information to build up a picture of the object. But Radon's theorem had referred to an infinite number of mathematically precise projections; what Hounsfield was thinking about was technically feasible methods of approximation. Pictures could be stored in computer memories in the form of two-dimensional arrays of numbers, each number representing the density (for example x-ray absorption density) in one small element of the picture (called a pixel). Hounsfield's approach was to depend vitally on his knowledge of computers.

In the few months it took Hounsfield to conceive the basic idea of computed tomographic scanning,[5] EMI management was considering the implications of the idea. Süsskind points out that the firm was very aware of its lack of experience with medical markets and of the difficulties it faced in assessing the feasibility of the project.[6] It is easy to see that existing x-ray suppliers would not have welcomed EMI's trespassing in their market even without invoking notions of "organizational field." EMI managers were well aware that the structure of the medical devices market, dominated to a greater extent by a few large firms than the computer market, would exacerbate the problems of entry. On the other hand, if their product was good, and original enough, it could provide a foothold in a potentially lucrative field. Hounsfield's superior, the director of EMI's Central Research Laboratories, was also a member of the board of EMI's Electronic and Industrial Operations (E & IO) division. He eventually decided to recommend Hounsfield's project for funding from a special pool of money that was reserved for more speculative research. Hounsfield's chief, Len Broadway, argued that they could easily sell ten instruments—four in Britain and six abroad—and thereby not only recover the R&D expenditure but at the same time obtain that foothold in the electromedical market. Süsskind quotes from his interview with Broadway: "As I saw it, the first essential question to answer, as far as the medical application was concerned, was whether an adequate signal/noise ratio could be obtained to show up the tissue structure of the patient without exposing him to an overdose of radiation. Work was therefore concentrated on this problem and Hounsfield was faced with the formidable task of developing the computer and getting the experimental equipment designed and functioning."

When the first allocation of funds was exhausted, the E & IO board refused additional support for Broadway without more concrete indications as to the market possibilities of the envisaged instrument. How would the clinical utility of the scanner be established? Contacts with radiologists were becoming essential, and these the firm lacked. At this point Broadway turned for help to the Department of Health and Social Security (DHSS). Collaboration between the firm and radiologists was going to be organized, quite unlike the chance encounter that led Donald to Kelvin Hughes or Lloyd Williams to Smiths Industries. Moreover the formal collaborative structure that ensued also provided a framework for negotiation over the sort of instrument to be developed. Starting from the assumption that the new device was to serve radiological purposes, negotiation would take place over "which purposes." In discussion with the ministry's technical staff, Broadway suggested that the device they had in mind, since it would detect tumors only one millimeter across, would be perfect for a mass screening program. The time in a sense was ripe. The period of particular fear regarding safety, in which use of x-ray had declined, was over, and there was considerable discussion of the benefits of breast and other screening programs. But this particular application did not appeal to DHSS, which looked at matters somewhat differently. Since a mass screening program entailed the purchase of large numbers of instruments, cost was necessarily an important consideration in decision making.[7] The new scanner was obviously going to be expensive. Out of these discussions emerged the first arguments for concentrating on scanning the *brain*. Departmental representatives pointed to a special need for a less invasive and risky procedure than those presently available (which had of course been Oldendorf's motivation). Almost any improvement would be of value, and if EMI could get their instrument to detect abnormalities in the brain about five millimeters in size, perhaps DHSS would be able to provide funding.

Hounsfield started work, using not x-rays but a weak source of gamma rays and a plastic water-filled phantom. Fixing his phantom on an old lathe, Hounsfield set about finding a solution to the problem of reconstructing its image from the projections produced by the beam of gamma rays. There are various mathematical procedures for reconstructing an image from its projections, and Hounsfield chose to work with a method known as the Algebraic Reconstruction Technique, or ART, algorithm. The basic principle is an iterative one: it relies on the power of the computer. An initial estimate of the two-dimensional array of numbers by which the reconstructed picture can be represented is first made. As experimental data for a given ray path is received, the computer compares that data with its estimate. The difference it detects is then divided among all the pixels intersected by that ray, so that estimates for each of these are changed. That of course has implications for all the other rays that intersect the original one, and thus each is modified. This method of successive approximation sounds somewhat like starting a crossword puzzle with a guess at the overall solution and then modifying it as "experimental data" are derived from the clues. It can be shown mathematically that if these corrections are made over and over again a good reconstruction will result. There are ways of improving the ART algorithm, through the introduction of "constraints" (Gordon, Herman, and Johnson 1975), but at this stage Hounsfield was not thinking about that.[8] The original scanner that Hounsfield developed made use of the "unconstrained ART algorithm."

Empirically, Hounsfield was trying to build up the image of his test object through gradually traversing it (taking a set of path readings along each traverse) and then rotating it to yield another set of traverse paths. The initial gamma ray source was so weak that generating the necessary data took nine days! Nevertheless, these data, fed into a computer and thence to a cathode ray tube, produced a fuzzy though definite picture. In January 1969 DHSS officials, including a radiologist, came to visit Hounsfield's laboratory. They were sufficiently impressed for DHSS to agree to back the construction of one or more machines and to arrange for clinical trials, provided it could be shown to work using x-rays and biological specimens. Each side would put in £5000 ($17,500).

During 1969 the system was gradually improved. The structures examined became correspondingly more complex, with animal brains (using skulls obtained from abattoirs) replacing physical phantoms. Eventually Hounsfield managed to obtain a perfect specimen of human pathology from the Royal Marsden Hospital— a suitably mounted horizontal slice of the brain of a patient who had died of a tumor. The results were remarkable, and on the strength of them EMI and DHSS signed another agreement in January 1970. At this point it was finally decided to develop the instrument in relation to the practice of neuroradiology. In addition to the argument from clinical need, made previously, a second set of (technical) arguments also played a part. Brain imaging did not pose problems of involuntary movements (heart beat for example, or twitching of muscles). Since scanning took several minutes these would be problematic in imaging many other areas of the body.

In planning to conduct tests on living patients, the question of selecting a hospital arose. Eventually a smallish neurological and neurosurgical hospital on the outskirts of London, Atkinson Morley's, was chosen. A leading neuroradiologist, James Ambrose, worked there, and he would collaborate with Hounsfield in further development of the scanner. At the same time DHSS set up an advisory team to oversee the project. So far as the hospital was concerned, the project was shrouded in secrecy; only Ambrose and the hospital administrator knew what the connection to the x-ray department actually was. There was nothing here of the humorous skepticism on the part of professional colleagues with which Donald had contended. Commercial interests had become more salient and the curiosity of possible competitors must not be aroused.

Süsskind's detailed account (1981) clearly shows how the stakes were gradually increasing. By mid 1971 EMI management was convinced that if they were to keep their lead, they would have to begin constructing further machines before assessment of the first prototype was even completed. DHSS began to realize that adequately backing the project would certainly take on the order of £250,000—a substantial part of the R&D budget. At this point, as Süsskind puts it, "DHSS made a courageous decision": to purchase one prototype and underwrite the construction costs for four more (p. 59). When an agreement to this effect was signed in July 1971, business minds were of necessity moving further ahead. What if the prototypes were a success? What should EMI do? For some months, Süsskind points out, debate raged. "Should EMI enter upon quantity manufacture and marketing of a product involving clinical radiological equipment and computers, two fields in which the firm had next to no expertise; or should it license others—perhaps an American company—to do it instead, and sit back to collect royalties?" (Süsskind, 1981)

It was a new technical director of EMI, John Powell, previously with Texas Instruments, who turned the debate in favor of going it alone. Powell joined the firm only in October 1971 and carried out an assessment of the firm's electronics business and R&D. The strategy he recommended to the board involved greater commitment to a more limited range of civilian electronic projects, and the CT scanner was an example of the kind of project he had in mind. Powell became the CT scanner's "product champion." "He . . . had decided that the CT scanner had tremendous potential for creating a market of which the company could conquer and retain a substantial share. But first he had to persuade his colleagues, and then the board. A multimillion pound investment was at stake—money that would not be available for other purposes." (Süsskind, 1981)

The implications of being new to the x-ray business were becoming apparent. Factories would be needed, so would experienced design and manufacturing engineers, and a sales staff able to sell to hospitals. EMI did not have these requirements, though it did have money. Powell argued that relying on the patent was risky. The patent would be assailed, there would be protracted litigation, and EMI would lose its head start.[9] Far better to plunge straight in, for it would take any competing firm a good two years to catch up.

Powell's arguments finally prevailed, and he was himself put in charge of planning a marketing strategy. It was immediately clear that a marketing campaign would have to focus on the United States: the scale of the market, the funds available for equipment, the commercial nature of health care in the United States all pointed to the United States as the crucial market.[10]

In April 1972 the new instrument was announced at a press conference, and over the course of the following months "and especially after the complete specification had been issued on 2 August 1972—EMI was overwhelmed by inquiries and approaches from every major manufacturer and supplier of medical electronics equipment, with the notable exception of . . . Siemens" (pp. 61–62). In November 1972, after careful preparation by Powell's team, Ambrose and Hounsfield went with the scanner to the principal annual radiological meeting in the United States, of the Radiological Society of North America (RSNA). Almost everyone present came to hear Ambrose's talk; the audience seems to have numbered several thousand, and the EMI stand was overwhelmed. It seemed clear that there would be plenty of buyers, at least in North America.

In 1973, in a paper published in the *British Journal of Radiology*, Hounsfield set out the ideas that had led him to the notion of the scanner and the principles on which it worked (see figure 5.1). The essential question for him, having once turned to the production of

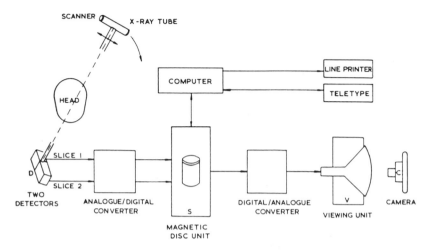

Figure 5.1
Block diagram of Hounsfield's prototype CT scanner. From Hounsfield 1973, 1017. Reprinted by permission of The British Institute of Radiology.

x-ray images, had been how more *information* might be obtained from an x-ray scan. Here is how Hounsfield presented the objectives of his own research:

For many years past, X-ray techniques have been developed along the same lines, namely, the recording on photographic film of the shadow of the object to be viewed. Recently, it has been realized that this is not the most efficient method of utilizing all the information that can be obtained from the X-ray beam. Oldendorff (1961) carried out experiments based on principles similar to those described here, but it was not then fully realized that very high efficiencies could be achieved and so, picture reconstruction techniques were not fully developed.

As the exposure of the patient to X-rays must be restricted, there is an upper limit to the number of photons that may be passed through the body during the examination, and so to the amount of information that can be obtained. It is, therefore, of great importance that the method of examination ensures that all the information obtained is fully utilized and interpreted with maximum efficiency. (Hounsfield 1973)

Hounsfield's scanner worked as follows. The patient's head was placed in a tightly fitting rubber cap, which formed one side of a plastic box filled with water. Absorption of x-rays by water is approximately the same as that by the head, so that by excluding air the range of readings that would be obtained—and processed—was thereby reduced. Replacement of air around the head by water thus served to reduce the calculations the computer would have to make. A parallel (collimated) beam of x-ray was passed through and allowed to impact on a sodium iodide crystal detector. This works because sodium iodide scintillates (emits photons of visible light) when struck by x-ray photons. The light emitted by the crystal was measured by a photomultiplier tube, itself connected to the computer. Source and detector were then moved along linearly. Information was collected at 160 points along this line, or "scan pass," and the x-ray projection was made up of these 160 readings (figure 5.2). The 160 readings for the projection were stored in the computer's memory, and the entire unit (source and detector) rotated through one degree around the patient's head. Another 160 readings were taken and the unit was rotated again, through a complete semicircle, thus giving 160 × 180 readings in all. This took five minutes (figure 5.3). The computer took a few more minutes to process the information from these 28,800 readings, which it could then present either as an 80 × 80 pixel array of x-ray density values or reconstructed as a picture.

Following Hounsfield's paper in the *Journal* was one from Ambrose, describing use of the new instrument and recounting his own first experiences with it (Ambrose 1973). Radiographers would have to become accustomed to using the instrument which, because it collects data in numerical form, is different from operating a traditional x-ray machine. He explained the four steps involved in its use. First, the patient has to be positioned in the water-filled rubber cap which must fit tightly on the head. Second, the plane being scanned has to be identified. That having been done, movement of the mechanical scanner has to be started from the console. The total run takes four minutes (the time taken for the x-ray tube to rotate through 180 degrees and to make 180 linear scanning motions), and for this length of time the patient must keep perfectly still. Finally, the radiographer must operate the viewer console after the picture has been processed. Ambrose pointed out that the information obtained by x-raying with the new machine could be displayed either quantitatively (showing absorption coefficients relating to the tissue examined) or "pictorially" (transformations of this quantitative data) (see figures 5.4 and 5.5). Much of the paper was devoted to displaying the relations of each of these forms of representation to the normal and abnormal anatomy of the brain. Ambrose had studied small numbers of patients showing various

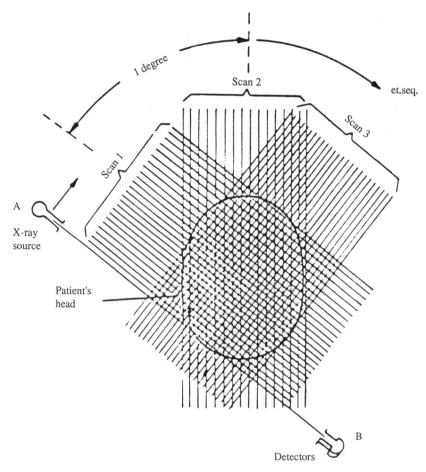

Figure 5.2
The scanning sequence of Hounsfield's prototype CT scanner. From
Hounsfield 1973, 1018. Reprinted by permission of The British Institute of
Radiology.

abnormalities of the brain regularly encountered by
neuroradiologists, and representations of these were displayed.
The paper finally speculated on the possible place the technique
might assume within neuroradiology and on the possibilities of its
improvement. Ambrose's first CT paper thus showed him attempting not only to prove that the new technology would be of use to
neuroradiologists but at the same time beginning to derive a future
research agenda from this assumption. So far as the first of these
objectives was concerned, the paper was in effect presenting legiti-

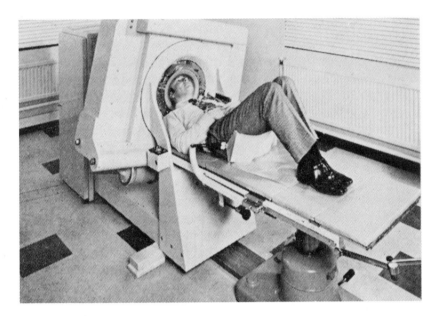

Figure 5.3
Hounsfield's prototype CT scanner, showing the patient in position. From
Hounsfield 1973, 1018. Reprinted by permission of The British Institute of
Radiology.

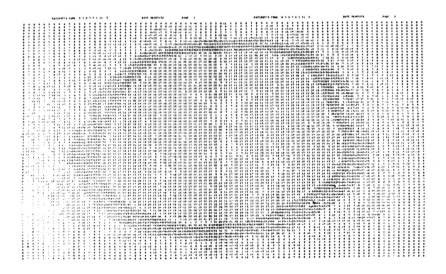

Figure 5.4
Normal CT scan in the form of a computer print-out of absorption coefficients
of matrix elements. A section 1.3 cm thick has been scanned. From Ambrose
1973, 1023. Reprinted by permission of The British Institute of Radiology.

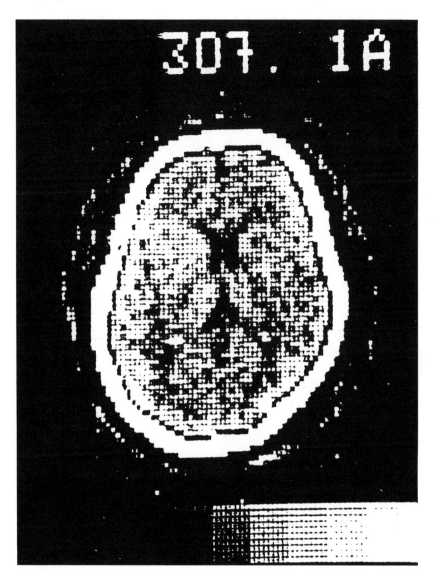

Figure 5.5
Polaroid picture of the scan shown in figure 5.4. High-absorption coefficients appear here as white areas. From Ambrose 1973, 1024. Reprinted by permission of The British Institute of Radiology.

mation for a community that had already become wildly enthusiastic—as the reception at the RSNA meeting and the stream of inquiries show.

Although at the end of 1972 EMI had no firm orders beyond the five machines being constructed under the agreement with DHSS, it was necessary to start considering the implications of production. Based on his interview with Powell, Süsskind reports that EMI reckoned that it might expect to sell fifty machines in the first year of production. That could serve as a planning target on the basis of which the necessary level of investment to start production could be calculated. It turned out that some £6 million ($20 million) would have to be invested in production, despite subcontracting some elements, before the first cash would start coming in.

As the exploratory period ended, the wild enthusiasm with which CT scanning was awaited is in striking contrast to the cautious optimism that infrared thermography had generated ten years before. Thermography in 1963 had not been in any way so glamorous, but more important it had been unclear *whose* instrument it was to be. By contrast, the development of CT scanning had begun within a framework of orderly negotiation between EMI and the radiologists, established by the Department of Health and Social Security. As a result, and in striking contrast to both the previous case studies, there was no doubt that use was to fall somewhere within the practice of radiology; negotiation hinged over where precisely. Ultimately, on the basis of considerations of need, cost, and technical feasibility, it was decided that the scanner be developed for purposes of brain imaging. But while for radiologists the possibilities inherent in the CT scanner generated only enthusiasm, for the manufacturers supplying their equipment the situation was potentially threatening.

The Development Phase of the CT Scanner, 1973–1975

Consequent on a more or less defined function within radiological practice was a more or less defined potential market. The consequence of *this*, in turn, was a potential threat to existing products and hence to existing producers. If this device really presaged a technological new frontier for radiology, then it would be crucial for existing suppliers to be at that frontier were they to preserve their reputations. Their determined attempts to catch up and

(ultimately) recapture the radiological market gave the development phase of CT scanning a *commercial* dynamic lacking from
either of the other techniques at this stage of their careers.

The development period was extremely short and it was a period
of intense activity. The original Hounsfield and Ambrose papers
appeared in late 1973. We have seen that radiologists were immediately enthusiastic. The machine was expensive, selling for $310,000
in 1973 and rising to $360,000 in 1974 (Office of Technology
Assessment 1978). Nevertheless by July 1975 EMI had already
installed over 120 scanners and by the end of the year had orders for
over 400 more. Meanwhile, though EMI had the bulk of the world
market, competition was building. By late 1975 no fewer than ten
companies were offering CT scanners for sale, although only three
(EMI, Ohio Nuclear, and Pfizer) were actually delivering (Süsskind
1981, 69). What happened in these three years of intense activity was
not simply the replication of EMI's design (though EMI did enter
infringement of patent litigation against the next three firms to
bring scanners onto the market). The development period was one
of *continuous* development, marked by an enormous rate of technical change, in which the scan time was reduced from minutes to
seconds and the applicability of the technique extended from the
head to the whole body. At the same time—thus during the
development period—radiologists were beginning to anticipate
how the new technology might be integrated in their professional
practice. The intensity and the variety of activity is reflected in the
research and development work to which CT scanning was subject
in this period.

The next five EMI prototypes, after that at Atkinson Morley's,
were installed in the course of 1973. One was located at the National
Hospital for Nervous Diseases (NHND) in London, one each in
Glasgow and Manchester, and two went to leading institutions in
the United States: the Mayo Clinic and Massachusetts General
Hospital.

These groups naturally had a head start in accumulating clinical
experience with the new scanner, in publishing their results, and
hence in adding to their scientific status. At Atkinson Morley's the
new instrument was in full diagnostic use after one year and as early
as 1974 Ambrose was able to report on a clinical series of 650
patients (Paxton and Ambrose 1974). Shortly afterward, in an
article in *Brain*, he set about assessing "not only the accuracy of the

CAT scanner as a screening technique but also the degree of confidence which can be placed in the definitive diagnosis made by this means" (Ambrose et al. 1975). By this time some improvement to the design of the EMI scanner had already been made. Replacement of the 80 × 80 pixel array by 160 × 160 produced a considerable improvement in picture definition (though the greater number of calculations placed new demands on computing). The NHND group, writing in the *Lancet*, were extremely enthusiastic, nor could they resist speculation regarding eventual integration of the scanner into neuroradiology. Of course radiographers would need to be properly trained if good pictures were to be obtained and adequately interpreted. Nevertheless, "The EMI scanner clearly possesses very considerable advantages. It is neither uncomfortable nor hazardous and can be readily used for outpatients. A positive diagnosis can be obtained in a very high percentage of patients as it records much more definite and detailed information than can be obtained from other screening procedures" (Gawler et al. 1974).

The clinical utility and accuracy of diagnosis achievable with the scanner were not the only issues being addressed. Studies were already anticipating not only accommodation of the new technique into diagnostic practice but, at the same time, the possibilities of technical improvement. Research in a number of centers differed specifically in terms of the *way* in which the technique as such was being problematized, supporting the view that we need to understand research agendas not in terms of the practice of clinical radiology alone but in relation to more complex sociocognitive structures.

The giants of the x-ray industry, potentially threatened in their core business by an outside firm, were of course obliged to respond. So they did, with determined efforts to get to the forefront of the field. Economic reasoning suggests that a rapidly growing market is a powerful incentive to invest in R&D aimed at product improvement.[11] Given the rapid development in the CT market between 1973 and 1975, we would expect a surge in industrial research and development activity. Moreover raising the R & D stakes is a means by which large firms, with vast resources, can try to squeeze out smaller firms unable to match their investments. This would soon follow.

We can see how the CT scanner was also leading to new problematizations for academic scientists associated in some way with radiology. At the Mayo Clinic a line of research "designed to realize the full potential of CT" had been started. One aspect of this related to "the augmentation of imaging and display capabilities and the extraction of maximal information from CT scan data" (Baker 1976). "Had Hounsfield and Ambrose been correct in their preference for the pictorial representation?" asked the Mayo scientists. True, radiologists were accustomed to looking at pictures, but perhaps information was being lost in converting the digital data into an image. "Is there more information on the numerical print out sheet than can be displayed and perceived on the cathode ray tube?" (Reese et al. 1975). By having the computer provide indicators of abnormality derived directly from the data fed into it, it was possible to problematize clinicians' observations. The diagnostic assessments made by clinicians viewing pictures could be compared with the computer's reading of the data. Mayo scientists carried out other fundamental studies, including problems associated with imaging objects in periodic motion, such as the heart (see Gordon, Herman, and Johnson 1975).

In other research groups different problems suggested themselves. Thus in the course of 1973 R. S. Ledley, professor of physiology, biophysics and radiology at the Georgetown University Medical School in Washington, D.C., set about developing a CT scanner. Ledley and his group had done a lot of work on pattern recognition in medicine, in fields as diverse as automatic chromosome analysis, analysis of Pap smears, thermogram analysis and echocardiogram analysis (Ledley 1974). Their expertise was thus just right. Ledley has explained how the idea of working on the CT scanner arose out of a local response to EMI's innovation: "The Georgetown University Hospital was interested in obtaining such a computerized tomograph but could not afford the $400,000–500,000 cost of financing the purchase of the EMI-machine. So the hospital asked me if my group could devote our efforts to building a computerized tomograph that would cost half as much, that would be operational in 6 months, and which, in addition, would be capable of scanning anywhere on the whole body" (Ledley 1974).

According to his own account Ledley thus set to work with the intention of building a relatively inexpensive whole body scanner. He soon concluded that for a whole body CT scanner it would be better to avoid use of the water bag which Hounsfield had introduced as an equilibrating medium. A much larger range of absorption values would therefore need to be handled. Ledley's other principal concern was motion. A whole body scanner would have to be relatively insensitive to periodic motions such as breathing and heartbeat. Ledley's group made rapid progress, and what they called the ACTA (standing for Automatic Computerized Transverse Axial) scanner, was operational by early 1974. Their very first picture was of a navel orange placed within a skull floating in a bowl of water, made on 11 January 1974 (Ledley et al. 1974a). The picture clearly shows the orange floating in the water that flowed into the skull. Though there are differences in design, the basic principles of the ACTA scanner were not greatly different from Hounsfield's EMI scanner. Like the EMI scanner, Ledley's ACTA scanner also formed its (160 × 160 pixel) image on the basis of 160 projections taken in each of 180 one-degree rotations. Its imaging time for the head was comparable. Ledley did choose a different computer reconstruction algorithm, closer to the mathematical method Cormack had discussed. The procedure made use of a theorem that relates what is called the "Fourier transform" of a one-dimensional projection to the Fourier transform of the two-dimensional projection of the same object.[12] By taking the "inverse transform" of the image built up in Fourier space from the 180 projections, one can generate a real image. So far as differences in design are concerned, the ACTA made use of three displays: a 19-inch color TV set and two 9-inch black and white sets, one of which is used only for photography and the other for viewing. Because of the variety of organ sizes it needed to accommodate, the ACTA had a choice of two scan passes, one 8.5 inches for the head and neck, one 19 inches for the thorax, abdomen, and pelvis. The first prototype was produced at Georgetown University, and patient studies were done—in fact largely neuroradiological studies—in conjunction with the university's department of radiology. Ledley's account of the celebrations following his success suggests the institutional significance work of this kind can have for a hospital. "Bottles of champagne suddenly materialized out of our neurosurgeon's desk drawer. The Chairman of the Physiology Department and Dean of the Medical School, the Medical Director and Administrator of the Hospital, the Vice-President for Medical Affairs, and all our friends showed up" (Ledley 1974).

At first glance this might be construed as a straightforward example of replication, akin to the attempts at building working TEA lasers Collins describes (Collins 1974). However the powerful influence of commercial and professional interests mean that what is at stake here is not simply reproducing the CAT effect. Both patent restrictions and the need to demonstrate advantage over competing equipment mean that basic principles (archetypes) have to be built into *different* working systems (types). It is worth noting how carefully Ledley tried to distinguish his machine from Hounsfield's in the publications announcing it. He repeatedly referred to a process of development lasting more than a decade (thus beginning with Oldendorff), and with all the basic work done by Cormack. Similarly, in referring to the potential uses of the technique, it is not CT scanning that is under discussion but ACTA scanning (Ledley, Wilson et al. 1974; Ledley, DiChiro et al. 1974). None of this prevented a later suit for patent infringement from EMI!

Ledley patented his ACTA scanner and set up a company, Digital Information Systems, to handle its commercialization. In fact the ACTA scanner was made and sold under license by the Pfizer Corporation, one of the world's largest drug companies. This came about as follows.[13] Pfizer had been engaged in trials of a pharmaceutical product, in which clinicians at Georgetown University Medical School were participating. A representative of the company, visiting Georgetown, heard about Ledley's new device and was invited to have a look at it. The meeting between Ledley and Pfizer corresponded to the interests of both parties. Ledley was looking for an effective way of getting his machine onto the market and was interested in licensing it. Pfizer, for their part, were interested in diversification possibilities, and a number of areas were at that time under active consideration. After a succession of Pfizer representatives had visited Ledley and his scanner, a deal was struck. In September 1975 Pfizer Medical Systems Inc. was set up to handle marketing of the ACTA scanner, and Pfizer thus became the second company to bring a CT scanner onto the commercial market.

Meanwhile scientists in Cleveland, Ohio, were also examining some of the issues involved in developing a whole body scanner. The possibility of whole body imaging, they write, "...has been the subject of considerable medical interest and the focal point of expanding research and development. Basic to these investigations is the determination of the attenuation characteristics of organs, other than the brain, in both normal and diseased states. Accordingly an investigation was undertaken to determine the attenuation characteristics" (Alfidi et al. 1975).

Having presented some of the attenuation characteristics that had been measured with the EMI scanner, the paper continued with consideration of the design requirements of an improved instrument, leading firmly from the biomedical laboratory toward the industrial plans. This Cleveland Clinic group was closely involved in the development of the Delta scanner, brought onto the market by Ohio Nuclear in June 1975.

Development work was already taking place under conditions of intense commercial competition. EMI was perforce keeping up with this rapid evolution, and was thus far maintaining its lead, though the cost was terrifying. "During the first five years after the initial commercial models were introduced, product development went ahead at a rate that made EMI management wonder whether the cost of any one design could ever be amortized. The reduction of the scan time to twenty seconds to make a whole-body scanner feasible had alone required considerable advances and redesign." (Süsskind 1981, 67). In March 1975 the international debut of the body scanner took place at the first radiologists' meeting devoted exclusively to CT scanning. EMI presented its new whole body scanner, the CT 5000, in which the original 80 × 80 matrix had now been replaced by a 320 × 320 matrix, which gave a much more detailed picture and in which scan time had been reduced to 20 seconds (with a 70 second "slow" option for reduced noise). This is an example of what is now referred to as a "second-generation scanner," with its fan-shaped beam and array of detectors. In September 1975 research prototypes of the CT 5000 were placed at the Mayo Clinic and at the Mallinckrodt Institute of Radiology in St. Louis. By November 1975, forty orders for whole body scanners had been received. At the same time the improved brain scanner, the CT 1010 with its 160 ×160 matrix, had been developed. Despite the extra computing involved, the scan time had been reduced to 1 minute through changing the computational algorithm.

Ohio Nuclear (now a subsidiary of Technicare), Pfizer, and Siemens were the second, third, and fourth companies to come on the market with "first-generation" scanners. Ohio Nuclear entered the U.S. market in June 1975 with the Deltascan, a unit capable of scanning in 120 seconds. In 1976 EMI initiated litigation for breach of patent against them.

GE entered the CT market in late 1975 by entering into an exclusive marketing agreement with a small, private, California-based company, Neuroscan, that had a brain scanner under development (which became CT/N). They began production of a body scanner of their own design (CT/T) in 1976.

Diffusion and Accommodation, 1973–1978

In the cases both of ultrasound and of thermography the diffusion and accommodation phase involved long and complex negotiations over uses and professional control. Thermographic devices had come very rapidly on to the market, but as the diffusion phase began controversy over its utility within radiological practice arose. There is nevertheless one similarity between the example of the last chapter and the present case. In both cases the diffusion and accommodation phase began at an extremely early point in the career of the technique. Early clinical results were going to be vital in marketing the new device. In both cases physicians closely associated with the industrial development work very soon began to consider the uses they foresaw for it. In his first CT paper, published in December 1973, Ambrose began to speculate on the place the technique might take in neuroradiological practice:

It is possible that the method will not immediately reduce the present practice of cerebral arteriography or pneumoencephalography but . . . it will undoubtedly come to occupy an increasingly important position, and . . . will no doubt supersede contrast radiological procedures in the investigation of many patients.
The hope is, that the majority of patients requiring investigation for a neurological complaint will need to be subjected only to plain skull radiography, a radioisotope brain scan and a computerized transverse axial scan. . . . These investigations should provide sufficient information in respect of location, tissue abnormality and structural displacement . . . for the institution of definitive treatment. (Ambrose 1973)

The sequence of problematizations was given by radiological convention. As experience and clinical data accumulated, attention turned to such issues as, How effectively could the various conditions confronting the neuroradiologist be diagnosed? How did the results of CT scanning fit with results from other procedures? Under what circumstances did the technique not work optimally? Ambrose found very quickly, for example, that the results of CT scanning, combined with those of the other "simple" procedures he had recommended, could be used to determine exactly when one of the more invasive techniques might need to be used (Paxton and Ambrose 1974). In a subsequent paper Ambrose moved on quantitatively to assess the accuracy and the confidence with which specific pathologies could be diagnosed relative to findings obtained with other techniques. His results, and those being obtained with the other EMI prototypes, were extremely promising. Yet these very first papers sought to convey something more than a sense of promise alone. They already displayed a sense of concern with how the technique was to be used, with its integration into the broader diagnostic context. It is partly, as Ambrose pointed out, that information derived from that context would frequently be required in the interpretation of scans. Beyond that, however, careful use was essential because these early experiences would decide the future of the technique. Presenting their first results in *The Lancet* in 1974, the NHND group wrote,

The avoidance of artefacts and equivocal results requires meticulous attention to technique by an experienced radiographer. . . . It would be tragic for such a promising diagnostic method to acquire a reputation for inaccuracy because of carelessness in the interpretation of its results.
The very ease with which an EMI scan may be performed may well pose problems. . . . It would be unfortunate if the EMI scanner were . . . employed to examine a large number of patients in whom a proper clinical assessment makes the presence of a lesion unlikely. Such a development would be a wasteful use of a method of investigation which has enormous potential for positive diagnosis. (Gawler et al. 1974)

Commercial availability of the head scanner was thus *preceded* by published suggestions as to what it could do and the place it might reasonably take in practice. But far more important for the growth of the market than these modest claims was the climate of expectant anticipation with which radiologists at large awaited the CT scanner. They would brook no interference. When a *Lancet* editorial, in 1974, wondered whether neuroradiologists and their neurological and neurosurgical colleagues "because of their specialized viewpoint, are necessarily the best people to advise on the strategic deployment of the EMI scanner in the community as a whole," neuroradiologists were "scandalized" (Jennett 1986, 197).

There was little doubt that a considerable market existed. We saw that a spirit of commercial competition marked even the development phase of CT scanning, as firms offered devices for sale before they were even in production. As CT's early promise was borne out by growing clinical experience, competition became all the fiercer. Positive clinical experiences, and the rapidity with which they were transmitted back to manufacturers, was reflected in a continuous process of further improvement of the scanner. With the subsequent availability of the whole body scanner the stage was set for a vast collective attempt to establish the range of investigations to which it should be devoted. At the same time professionals were beginning to think about some of the broader implications of availability of the scanner for their practice. By the late seventies there was much talk in policy circles of the costs of health care and growing concern with enhancing the cost consciousness of physicians. As radiologists responded to this changing climate new sorts of questions regarding the integration of the scanner into practice, including the implications for other radiological procedures, began regularly to be posed.

From June 1973 until October 1974, the rate of installation of CT (head) scanners in the United States averaged less than 5 machines per month. By late 1974 there were some 45 scanners working in the country. During 1975/76 the rate of installation rose to 19 per month, with 475 scanners working by the end of 1976. Only manufacturing capacity limited growth, and EMI had a backlog of orders for 250 scanners at the end of 1976 (OTA 1978, 48). Growth continued through 1977, with 40 scanners installed in the average month (Banta 1984, 70), and at least 873 operational by November. Although legislation extending the authority of the Food and Drug Administration (FDA) to preapprove medical *devices* had been passed in 1976, the CT scanner had just escaped the more rigorous assessment for which the law provided. Certificate-of-Need Legislation, though in place in many states, had not yet affected the new device. Given the buoyancy of the market the extent of industrial interest is no cause for surprise. By the end of 1975 no fewer than twenty firms had developed or were developing CT scanners.

Radiologists fortunate enough to have been in on the ground floor continued with their assessments of the scanner's diagnostic utility, as this paper from the Mayo Clinic suggests. "More than 400 patients with a wide variety of clinical problems were examined during the first 4 months of use of this scanning unit. Almost all patients had, or were suspected of having, a mass lesion. Special emphasis was placed on patients who were likely to have a diagnosis confirmed by a surgical procedure or who had a relatively definite diagnosis established by another reliable diagnostic modality" (Sheedy et al. 1976).

Here, and in other work, the focus is on the questions, What does the CT scanner do? and How well does it do it? Radiological practice gives content and significance to these questions. In order to pose the second question, cases being counted have to be classified on a basis deemed authoritative: either a so-called gold standard (autopsy or biopsy) or, more usually in practice, "another reliable diagnostic modality." This is not only a methodological point. With what precisely is the new technique to be compared? What "reliable diagnostic modality" has the authority to validate its findings? In this sense the assumptions underlying clinical practice at any one time, the status of existing techniques within that practice, shape the kinds of studies through which first use and then accuracy of the new technique is established. Through studies of these kinds, reflecting the then-current state of radiological practice, clinical uses for the brain scanner and, later, for the whole body scanner were established.

Radiologists not among the fortunate few were hardly inclined to wait for the results of trials before they maneuvered to obtain CT scanners for their institutions. Their enthusiasm had previously provoked a critical *Lancet* editorial. In 1977 Creditor and Garrett, writing in the *New England Journal of Medicine*, took a different but no less critical line, pointing out that with an average price of half a million dollars, some $700 million would have been invested in CT scanners by the end of the year. They go on: "It would seem that an investment of this magnitude would require evidence of considerable potential marginal gain. There can be little doubt that computed tomography is a safe and effective modality for neurodiagnosis, but did the 13 clinical papers in the English-language literature by June 1975 justify the ordering of 100 units? Did those papers or the 40 clinical papers published by December 1975 provide evidence that the contributions added by computed tomography could justify the enormous investment?" (Creditor and Garrett 1977).

Creditor and Garrett were not of course claiming that questions of this sort were not being posed. The argument was rather that it would have been proper to await harder empirical evidence regarding the utility of the device before investing in it on any scale. But radiologists had their own reasons for not waiting, as did hospitals. Indeed, part of the urge to get hold of a scanner was precisely to be *able* to investigate its applications and utility. By early 1975 Mayo scientists had already been writing, "In the past few months, neuroradiologic practice in a number of centers throughout the world has been affected by the advent of computed tomography (CT) of the head. The full impact of this unique, noninvasive radiologic method of diagnosis will not be felt for some years; but early experiences seem to indicate that a general reassessment of our modes of practice may be in the offing" (Baker et al. 1975).

As the device came to take its place in the clinic, so that patients to be scanned were no longer selected according to a research protocol but according to normal processes of clinical decision making, physicians began to look back at the effects of the new technique on radiological practice. How and why did clinicians seek a CT scan? What effects did availability of the technique have on use of other techniques? What did the machine cost to run, and how intensively could it be operated? These questions, too, were being posed in scientific reports. Here we see the integration of the new device into clinical practice being problematized in a new way, not in relation to the individual diagnostic act but to that of the social and technical organization of diagnostic radiology.

Papers often move imperceptibly from accounting for what *was* done to suggesting what *should be* done: "Within the analysis of brain tumor patients a more substitutive role for CT scanning appears. Both EEGs and . . . were used less frequently and while no direct effect was documented for cerebral arteriograms, the overall volume of neurosurgery supports a more limited role" (Knaus et al. 1977).

The period 1975 to 1977 was the heyday of CT scanning, with the rate of installation in the United States starting to decline thereafter, as shown by the diffusion curve.[14] Purchased first by large university affiliated hospitals, CT scanners gradually spread into smaller institutions, "promoted aggressively by influential hospital-based physicians" (Banta 1984, 84). In line with expectation decision making, and the information required, began to change. Recall how in chapter 1 we had the notion of adoption of innovation gradually ceasing to be based on a reasoned assessment of the task at hand and becoming more a kind of "copying behavior". DiMaggio and Powell's analysis seems highly applicable here, with rules of behavior set by the established organizational field, or interorganizational structure. Though radiologists may act to secure the new technology for their hospitals, there is also evidence that "few . . . follow the professional literature on a new technology . . . when departments acquire new technologies most members will know little about the machine or its images" (Barley 1986). Crucial data gradually shift from the context of the diagnostic act to that of the overall operation of a diagnostic facility. By the mid seventies, with new cost-containment measures in place, the economics of new technology was naturally widely discussed in the medical community. As costs and financial benefits came to be an important consideration in hospital decision making, detailed information on experience with CT scanners became critical in purchase decisions.

In a 1976 article in the *American Journal of Roentgenology* Evens and Jost of the Mallinckrodt Institute of Radiology in St. Louis presented financial data regarding 98 of the 140 CT scanners installed in the United States as of January of that year (Evens and Jost 1976). EMI models accounted for 81 of the instruments, almost all head scanners. The paper provided detailed information on such issues as the type of funds with which scanners had been purchased, average space allocation, and the costs of maintenance (typically maintenance contracts with manufacturers cost around $25,000 per year). In full-time equivalents, the average personnel assigned to a CT scanner was 0.96 radiologists, 2.3 x-ray technologists, and 2.3 other personnel—a hefty personnel complement. Detailed information on depreciation and principles of charging was also given, as was a graph showing the number of patients needing to be scanned per week, at various charging levels, for break even and profit. Charging $150 for a scan, the average radiology department would need to examine 49 patients per week to cover its total costs. This distillation of the experience of other users was extremely valuable information for the hospital contemplating the investment of $300,000 to 400,000.

Despite its clinical success, early market forecasts for the CT scanner proved to have been wildly overoptimistic. The newly installed Carter administration was determined to do something about the rising costs of medical care. Süsskind quotes an EMI executive to the effect that "CT scanning—particularly body scanning—became a whipping-boy of these cost containment policies. Some strident critics claimed that it was a glaring example of a new technology attracting many millions of dollars . . . and yet it could not produce the evidence to prove that such a high level of interest . . . was justified" (Süsskind 1981, 72). By 1977/78 the market had ceased to grow significantly, but the place of the head scanner at least within neuroradiology had been secured.

Feedback: 1973 to the early 1980s

In writing about the development phase of CT scanning, I hinted at the impossibility of any clear demarcation from "feedback." So clear were the principles involved, so obvious the possibilities (both radiological and commercial) that the merest indication of what was in the offing was sufficient for competitors. Availability of the EMI scanner posed a challenge of considerable magnitude to existing industrial interests in production and innovation in x-ray instrumentation. Faced with the necessity of developing devices with which they could compete with EMI, major manufacturers drew in relevant expertise, and then increased the stakes in the R&D process. The existence of a clear market and the possibilities that offered led not only to defensive-offensive strategies on the part of existing x-ray manufacturers. Others too were drawn in. We saw how Ledley set about trying to build a cheaper instrument, capable of scanning anywhere in the body, in the course of 1973 and how this device became the basis of Pfizer's entry to the medical equipment market.

The improvements that had to be made to the first models were clear. Radiologists were demanding still faster imaging and still better resolution, and producers competed in trying to respond. The rate of improvement was relentless, and the necessary investments in development work formidable. As Dr. G. M. K. Hughes, then involved in a senior capacity in Pfizer's CT activities, put it, "Those of us who got in early were faced with the *necessity* of bringing out a new model each year" (my emphasis).[15] By the 1976 Radiological Society of North America convention, at which all manufacturers exhibited, the (British) weekly *New Scientist* was beginning to wonder if EMI would be able to hold on to its lead. "Probably the biggest new challenge will come from General Electric," wrote their correspondent, "whose strong name will give them credibility in the U.S. marketplace" (Kehoe 1976). The challenge that would be mounted by established industrial interests (GE, Philips, Siemens, and the rest) was becoming clear, and so was the general form of more advanced systems. Kehoe reported, "Hounsfield predicts that different scanning systems are likely to evolve, which could be faster. One such system, which has been designed in principle by a research team at the Mayo Clinic, uses multiple x-ray sources which are pulsed in phase. Leading manufacturers are believed to be examining this and other approaches to advanced CT scanning. Further developments may also be expected on the data processing side. . . . The trend is to manipulate the data, rather than to move the patient" (Kehoe 1976).

Because of the clearly "linear" direction of improvement it became usual to discuss models in terms of "generations." First, second, and third-generation scanners were all in existence by mid 1978 and are described in the OTA study of the scanner published in August (Office of Technology Assessment 1978). The increased speed of imaging, vital in the case of whole body scanning where patient motion could destroy the image, entailed significant redesign, and considerable work on the computer algorithms used in data processing. The OTA report, summarizes major differences between the three generations of scanner then available. First-generation scanners used a single source and single detector: moving in one-degree increments they took some five minutes to complete one scan. Second-generation scanners were much faster. They typically possessed either a single source producing a fan-shaped beam or multiple pencil beams and multiple detectors. The gantry moved in larger steps, with more information being collected at each, and scan time was reduced to between twenty seconds and two minutes. Third-generation scanners, on the market by 1978, used a fan-shaped beam of x-rays wide enough to encompass the entire object being scanned and a bank of detectors. With the need for lateral passes eliminated, the source and detectors could be moved around the object in a continuous sweep. Scans could typically be completed in five seconds (figure 5.6).

Clear indications regarding technical improvement were not the only signals being fed back to producers. By late 1978 Certificate-of-Need (CON) controls had started to bite, and the CT market in the United States was no longer growing. Moreover not only was the market now shared by numerous manufacturers, but there seemed to be no let up in the further investments in R&D needed to keep up. Some manufacturers were being forced to reconsider their positions. For EMI, the medical electronics division had been astonishingly profitable in 1976 and 1977.[16] In 1978 EMI announced its series 7000 scanners. Including five distinct models, the "7000" models were the results of a two-year development program which had cost the firm some 20 million pounds ($44 million). By 1978, however, sales and profits from CT scanners had collapsed.[17] By 1979 the parent company was itself in deep trouble. EMI saved itself by a merger with Thorn Electrical Industries, but in early 1980 Thorn EMI Ltd. announced that it was withdrawing from the field of medical electronics. Most of EMI Medical was sold to its chief rival in the CT scanning field, GE. Godfrey Hounsfield, who had started it all little more than a decade before, would not be forgotten

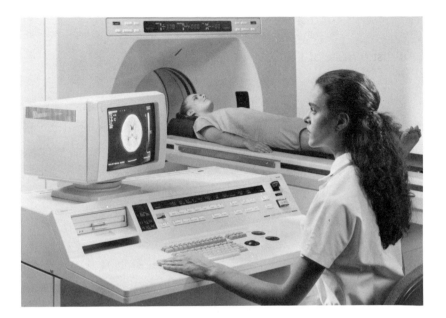

Figure 5.6
A modern Tomoscan in use. Reprinted by permission of Philips Medical Systems.

however. In October 1979, together with Cormack, he received the Nobel Prize for medicine or physiology.

Meanwhile a more general shakeout in the industry had been taking place. By 1978 GE and Technicare (Johnson & Johnson) had over 70 percent of the U.S. CT scanner market. Of the 15 or so firms that had been marketing CT scanners worldwide, a number of others also decided to withdraw. In addition to Thorn EMI, they included Searle, Syntex, and Pfizer. In the case of Pfizer this entailed a fundamental assessment of the company's activities. With sales of $2 billion in 1977, Pfizer was a major force in the world of pharmaceuticals. It became clear that maintaining a position in the CT market would have major implications for the company. As Dr. Hughes put it to me, so far as diagnostic imaging is concerned, you either had to become a fully fledged radiology company—or you got out. Pfizer decided to cut their substantial losses and get out of the imaging market. The CT scanner, unlike either of the previous technologies analyzed here, had from its origins been developed in relation to the practice of radiology. A corresponding industrial structuring had taken place, and the interorganizational structure reasserted its control of professional and manufacturing activities related to x-ray technology.

By 1982, although further development work was being done—notably by GE and Johnson & Johnson—a review of the diagnostic imaging industry took the view that "the major improvements in image quality have already been made with the present technology," so that "only small incremental improvements will be made in the future, given the cost limitations and dose restrictions" (Hamilton 1982, 82).

Unlike either of the previous technologies, the CT scanner was fundamentally based on computer technology. As advances in semiconductor technology made it more widely available, the computer offered the prospect of dealing with image reconstruction problems that had arisen in various areas of science (e.g., radio astronomy, x-ray crystallography). The well-known radiological problem of imaging "in depth" could also be conceived as of that sort.

Among those who saw the relevance of expertise in this area of computing for x-ray diagnosis was an electrical engineer working in EMI's Central Research Laboratory. Lacking experience with the medical market, EMI was unable to assess the project he proposed. Through the mediation of the Department of Health and Social Security in London a framework for interaction with the radiological profession was established. The subsequent "negotiation" established to *which area of radiological practice* the proposed innovation was to be addressed. As CT scanning came to be constituted as a means of performing neuroradiological diagnosis, the market it would have to occupy was simultaneously defined. In other words from an early stage it was quite clear which existing commercial interests were (potentially) threatened. This is in striking contrast to the previous case studies. Swelling professional enthusiasm made this an area of vital importance both for existing producers of x-ray equipment—potentially threatened—and for firms looking to enter the imaging market. A commercial dynamic quite unparalleled in medical imaging came to dominate.

A rapidly growing market stimulated vast investment in R&D. One generation of scanner followed another, and the mid 1970s were a period of rapid adoption of the new technology. Producers could scarcely keep up with demand. The CT scanner became something any self-respecting department of radiology simply "had" to have: a good example of the "institutional isomorphism" characterized by DiMaggio and Powell.

By the end of the 1970s, however, this market had been transformed. To health authorities the CT scanner began to symbolize health budgets that were coming to be seen as uncontrollable. New policy measures began to bite. Shrinking sales, combined with the seemingly inexorable R&D costs of staying abreast, led many firms to drop out of a market that was henceforth dominated by the x-ray supply industry. In any event a new and more glamorous claim on R&D funds available for investment in diagnostic imaging apparatus had arisen. This was nuclear magnetic resonance imaging.

6

The Constitution of Magnetic Resonance Imaging

Introduction

The technique now generally known as magnetic resonance imaging (MRI), like CT scanning, is inconceivable without modern computer technology. Here too, images are *reconstructed* from projections and can be manipulated in a variety of ways. Unlike the CT scanner, however, its image does not depend upon the use of x-rays. Moreover, and here the parallel with thermography more readily comes to mind, the images formed are not self-evidently derived from the *structure* of the body being imaged. Recall that the novelty of thermography was initially held to inhere in the fact that it probed physiological process, not morphology. We shall see that something similar can be said of magnetic resonance imaging. Devotees of thermography sometimes argue that the point was either lost on or ignored by radiologists, who evaluated the technique in their own, essentially morphological terms. That issue will arise here too.

In 1924 Wolfgang Pauli suggested that atomic nuclei having odd numbers of protons, or neutrons, or both, would behave as tiny magnets. Atoms having this property should include frequently occurring isotopes (forms) of many common elements such as hydrogen (^1H), sodium (^{23}Na), and phosphorus (^{31}P). Crucial experiments were carried out in the 1930s (Laitanen and Ewing 1977, 204), although it was to be some years before the first direct observation of the nuclear magnetic resonance (NMR) phenomenon took place. Once again wartime experience, specifically with radar, proved important in designing the equipment needed to observe the effect in bulk materials. In late 1945 Bloch at Stanford and Purcell at Harvard independently performed the nuclear magnetic resonance experiment, work for which they shared the 1952 Nobel Prize in physics.

The experiment depends upon the fact that when atomic nuclei having a magnetic moment are placed in a magnetic field, they precess about the direction of the magnetic field with a frequency that depends on the strength of the magnetic field applied, but which usually falls in the shortwave radio region. A useful analogy for understanding this is the gyroscope. When a spinning gyroscope is tipped from its vertical axis, it rotates about this axis in the shape of a cone. In the magnetic resonance apparatus a set of coils at right angles to the main magnetic field produce a radio signal. If this signal is arranged to be at the right frequency, atoms in the sample will absorb energy; this is the phenomenon known as *resonance.* Some of the atomic nuclei will start to precess in step with each other. When the signal is turned off, the energy absorbed is lost. A small voltage is created, which decays as the atoms return to equilibrium with the magnetic field. The magnitude of the voltage, called the *spin density,* relates to the concentration of nuclei (atoms) in the sample. Energized nuclei lose their energy through two distinct "relaxation" processes, which lead to the definition of two characteristic parameters called T_1 and T_2. T_1 is called the *spin-lattice relaxation time.* Representing thermal transfer of energy to the environment, it is the time required for nuclei to realign with the static magnetic field. T_2, the *spin-spin relaxation time,* relates to the fact that the precessing nuclei eventually get out of phase with each other through transfer of energy to other nuclei that were not energized, thus ultimately canceling out the net voltage effect. In 1949/50 it was discovered that the same atom (copper, fluorine, hydrogen, and nitrogen were all studied) when contained in different chemical compounds had different resonance frequencies. This finding became the basis for the very important role NMR acquired in the determination of molecular structures. Commercial instrumentation gradually became available, from the late 1950s, and NMR eventually acquired a major role in the chemical laboratory.

It was to be another fifteen years before the notion of NMR as the possible basis for a diagnostic imaging technique emerged. It eventually did so just when both radiologists and industry were preoccupied with the possibilities of CT scanning. In the mid 1970s the idea that a new medical imaging technology could generate vast profits took firm root; however, the years following showed that these profits were not easily won. To compete successfully in that particular market entailed more than having a good product to sell. Recall Pfizer's CT experience, summed up by Dr. Hughes, to the effect that the field admitted of no half-measures: one either became a full-fledged radiology company—or one got out. Contemplating what seemed to be the possibilities of magnetic resonance, firms were very conscious of their own experience with CT development and the CT market. Pfizer, for example, resolving *not* to become a "full-fledged radiology company," abandoned both CT manufacture and its MRI development program. Philips, conscious of having entered late into CT scanning, resolved not to make the same mistake twice. Meanwhile development costs rose inexorably, while the regulatory climate—reflecting financial curbs on public health care expenditure—became ever tougher. Radiologists, however, could look with unrestrained enthusiasm toward the future. Harnessing the computer to their traditional technology had shown them what a technological breakthrough could amount to; they were eager for more. Because of the preoccupation with CT scanning, it took time for these attitudes to crystallize. In that time—and this is crucial to understanding the emergence of MRI—a somewhat different set of expectations was also developing. Arising from application of established NMR laboratory techniques in biomedical research, these were not specific to radiologists. As in the case of thermography, the hint of a possible means of early cancer detection, emerging from this prior research, was powerfully to influence the way in which MRI was to be conceived.

In 1971 a paper appeared in *Science* with the arresting title, "Tumor detection by nuclear magnetic resonance." The author was Raymond Damadian, a physician and assistant professor of biophysics at Downstate Medical Center in New York (part of the State University of New York system). In the late 1960s Damadian had come to share a theory of the abnormal water structure of the cancer cell first put forward by Gilbert Ling. According to this theory, when a cell becomes cancerous, it loses its structure as well as the capacity to distinguish between sodium and potassium ions. It fills up with water. There ought to be a major difference in water structure between normal and cancerous cells. Through contacts made at a 1969 conference, Damadian was persuaded that he could usefully test this idea by looking at NMR spectra. In June 1970 he carried out his experiment. Damadian took six tumor-infected rats to the premises of a small NMR company, NMR Specialties, where he was able to put samples of cancerous and normal rat tissue into the NMR spectrometer. Results bore out his expectations. His study supported the hypothesis that "cancerous tissue has a lower degree of organization and less water structure than normal tissue." In the conclusion to his *Science* paper Damadian offered what may have seemed a farfetched speculation. "This technique may prove useful," he went on, "in the detection of malignant tumors," since NMR combines "many of the desirable features of an external probe for internal cancer" (Damadian 1971).

Damadian's article attracted some attention, and a good deal of skepticism, from biophysicists and NMR researchers. Early attempts at repeating his comparison of normal and tumorous tissue, largely successful, were motivated by a variety of interests. Smidt (Delft University of Technology) and Hollis (Johns Hopkins University), among others in the NMR community, were curious to test out this surprising claim for the NMR technique. Like Damadian himself, a few biomedical scientists, such as Hazlewood in Texas, were looking for support for their theory of an association between cancer and the structure of cell water. John Mallard, professor of medical physics at the University of Aberdeen, jumped to the same conclusion as Damadian and immediately began to think of the possibilities in clinical diagnosis.

The fact that Mallard made the same leap of faith as did Damadian is understandable in terms of his background, which involved earlier work on new diagnostic modalities. In 1953, at London's Hammersmith Hospital, Mallard had become involved in pioneering work on radioisotope scanning and had in fact constructed the first gamma camera in Britain.[1] In the early 1960s Mallard had begun investigation of the possibility of tissue characterization using another recent laboratory tool, electron spin resonance, or ESR.[2] When he moved to a new chair of medical physics in Aberdeen in 1966, Mallard continued to pursue these interests and some results were published (e.g., Hutchison, Foster, and Mallard 1971). However, the work subsequently ran into difficulties and was terminated. Shortly afterward, Damadian's 1971 paper appeared. Mallard recognized that Damadian had been doing what he had wanted to do, but using NMR in place of ESR. Immediately a Ph.D. student,

R. E. Gordon, was set to work building an NMR spectrometer and using it to measure tissue relaxation times.

By 1972, having with some difficulty obtained three-year funding from a skeptical National Institutes of Health (Kleinfield 1985, 67–68), Damadian had developed his ideas further. It seemed that a new and more meaningful way to *classify* tumors was at hand. He related, "Methods to date are mainly descriptive, relying chiefly on the appearance of tumor specimens in the light microscope. Apart from acknowledging the added power of discrimination implicit in the assignment of a number to an observation, it is worth noting that although a chemical reagent is the ultimate objective of a cure for cancer, study of this disease is largely devoid of a chemical data base and remains rooted in morphology" (Damadian et al. 1973).

With the aid of NMR, it became possible to think of "reclassifying tumors according to their chemistry rather than by their appearance in a light microscope," which might in itself have major implications for therapeutic progress. Moreover, Damadian had worked out how he could focus the NMR signal to examine one small region in the body. This problem of spatial localization was crucial if the NMR technique was to become a method of *scanning.* It would be essential to be able to take the signal from successive and known points in the body. Damadian hoped to do this by using a magnet producing a (nonhomogeneous) saddle-shaped field; in this way he could arrange that the conditions for resonance would be obeyed in only one small area. In March 1972 Damadian filed a patent application on his idea. He was more than ever convinced that NMR had already earned its place in the hospital pathology laboratory "as an adjunct to present methods of diagnosing malignant lesions."

However, in the years following, these ideas were challenged by various workers. It began to seem that T_1 elevation in tumor-bearing mice was not due to some characteristic feature of carcinogenesis, leading to changes in the structure of the tissue water, but simply to a higher water *content.* The problem, of course, was that even if cancer did lead to a higher water content, so did many other traumatic and inflammatory conditions. If this is what was being detected, then both the scientific claim (that something interesting about carcinogenesis was emerging) and the clinical claim (that a diagnostic test was at hand) were threatened. This was a difficult time for Damadian.[3]

The Exploratory Phase of Zeugmatography: 1973–1977

Damadian was not at that point thinking of diagnostic *imaging*. Although he believed he had established a connection between NMR parameters and cancer and had evolved the outline of a method for scanning the body, Damadian had not envisaged transforming the NMR signals into images. An NMR expert, the chemist Paul Lauterbur, is credited with making that connection. With the publication of his first paper on the subject, in 1973, the exploratory phase of magnetic resonance *imaging* opened.

Though it would take only four years to produce the imaging device, and then the image, which would signal transition to the development phase, MRI's exploratory phase was complex. It is not just that the origins of this imaging technique seem more than usually disconnected from radiological practice. As in the case of ultrasound, early investigators approached their work with a variety of notions of its possible medical relevance. The consequences of this diversity were to be quite different from the ultrasound case, largely for one simple reason. Many early workers were not physicians, so that their sense of MRI's medical applications would not flow from professional experience. In this case we do not find the independent constitution of the technique in distinct medical contexts that we encountered with ultrasound. What characterizes the exploratory period of magnetic resonance imaging is the gradual incorporation of medical goals into research initially rooted in physics (and in some cases in chemistry). As their research evolved under a new and unaccustomed discipline, many of these physicists and chemists were obliged to make certain important accommodations. As anatomists, physicians, and then the formal institutions of medical science became significant for their work, they found themselves in an increasingly unfamiliar—hybrid—world.

A professor of chemistry at the State University of New York at Stony Brook, Paul Lauterbur, used a modified NMR technique to generate two-dimensional representations, or images, of simple objects (Lauterbur 1973). Lauterbur had been working with NMR for many years, had published widely on the NMR spectra of organic molecules and on problems of stereochemistry, and was on the board of directors of NMR Specialties. Through that connection he may have heard something of the experiments that Damadian had carried out there.[4] The procedure Lauterbur announced in 1973 represented a significant departure from the NMR technique used in the chemistry laboratory. The essence of it is simple enough. In NMR spectroscopy as practiced by chemists, the sample examined is small and homogeneous (for example, 1 cc of liquid contained in a tiny tube). At the same time the magnetic field is as uniform as possible. The idea is to obtain an identical signal from each atomic nucleus within a given conformation. Lauterbur's idea involved deliberately making the magnetic field nonuniform. Superimposed on the uniform magnetic field is an additional field *gradient*, so that different parts of the specimen find themselves in different magnetic fields. They therefore have different resonance frequencies so that only parts of the specimen will resonate at the frequency of the applied radio field. Signals can then be obtained from locations within the sample, rather than from the sample as a whole. In this way it is possible to plot signal strength in relation to distance and thereby construct a one-dimensional map, or profile, of proton density along the direction of the applied magnetic gradient.[5] In his article, Lauterbur showed that it was possible to *rotate* the magnetic field gradient around the test object and *combine* the projections obtained at each position by means of a computer algorithm similar to those being used in CT scanning. This had been achieved with simple test objects, such as a tube of water. Lauterbur gave the name *zeugmatography* to his technique, from the Greek word *zeugma*, meaning "that which is used for joining." A number of possible applications were envisaged for the new technique, specifically including medical imaging. Lauterbur wrote, "The variations in water content and proton relaxation times among biological tissues should permit the generation of useful zeugmatographic images from the rather sharp water resonances of organisms, selectively picturing the various soft tissues and structures. A possible application of considerable interest at this time would be the *in vivo* study of malignant tumors, which have been shown to give proton nuclear magnetic resonance signals with much longer water spin-lattice relaxation times than those in the corresponding normal tissue (Wiseman et al. 1972)" (Lauterbur 1973).[6]

A number of workers immediately set out to do more or less what Lauterbur had done. "More or less" because different elements in Lauterbur's account were emphasized and problematized, reflecting different social and cognitive contexts. For some the suggested application was the major stimulus, while for others it was the imaging method (compare Amsterdamska and Leydesdorff 1989).

There were those, including notably Damadian and Mallard, who were already working on clinical applications of NMR. For them, Lauterbur's paper was something of a blow. Kleinfield quotes Damadian's enraged recollection: "Here was I talking about medical scanning and getting ridiculed and here was this guy standing up and saying that he had invented it. I was absolutely shocked. I couldn't believe it." (Kleinfield, 63). For John Mallard, Lauterbur's paper was also a blow, though a less personal one. Mallard explains that he too was on the way to imaging, but had not gotten to it when Lauterbur's paper appeared. A method was now at hand, and Mallard and his colleagues took immediate advantage of it to push rapidly ahead. From their department's own resources a new instrument was constructed with a working space of 3.2 cms (large enough to accommodate a mouse) and a 25 MHz permanent magnet. Using Lauterbur's method, they quickly obtained a crude image of a mouse (Hutchison, Mallard, and Goll 1974).

On the other hand there were NMR researchers whose interest was aroused less by the possibility of imaging malignant tumors than by the idea of NMR imaging as such. Replicating Lauterbur's work thus had a different significance for them. They were in the first instance motivated to see if they could do—or better still improve on—what Lauterbur had done. Two of these, for whom Lauterbur's paper was an intriguing challenge, were E. R. Andrew and P. Mansfield, both of the physics department of Nottingham University.

After wartime work in radar, Professor Andrew had devoted much of his scientific career to radiofrequency spectroscopy.[7] Andrew was intrigued by Lauterbur's paper, and when he and Waldo Hinshaw (an American postdoctoral fellow working in his laboratory) heard Lauterbur lecture, they decided to try NMR imaging for themselves. Right away, however, their enthusiasm was tempered by a feeling that "there might be better ways of doing it" that did not require the large signal handling and computing facilities needed by Lauterbur's method. Lauterbur's idea also connected closely with Mansfield's existing interests in physics. Having been working on the structures of solids, in 1973 he had published a method for forming one-dimensional projections of proton density (Mansfield and Grannell 1973). Though less ambitious than what Lauterbur proposed, Mansfield's method was nevertheless along the same lines.

The subsequent research programs of the various groups reflected these different starting points as well as the different social and cognitive contexts within which the groups worked. For the NMR physicists, Andrew and Mansfield, the starting point had been the technical question of improving Lauterbur's image reconstruction method. For Damadian and for Mallard, the aim—in view from the start—had always been connected with detection of cancer.

Andrew's group immediately set about making the modifications to their laboratory spectrometer needed to construct an image. Progress was rapid. By 1974 Hinshaw had conceived two possible ways of image formation that avoided complex data processing (Hinshaw 1974). These introduced a time dependence to the magnitude and direction of the field gradient. A method that seemed particularly simple used a set of three intersecting coils and a homogeneous magnetic field to define a single point at which there was zero magnetic field gradient. By altering the current in the various coils relative to each other, this point could be moved through the sample, and a filtering device could be used to take the signal from the chosen point alone (see figure 6.1).

Like Andrew and Hinshaw, Mansfield was looking to develop his own, better—and in his case specifically faster—method of image formation. For Mansfield too the problem was to find a means of selecting out the signal from a tiny element within the sample and sweeping this element across the sample, thus building up a picture element by element. Mansfield's method, as described in 1975, was briefly to "irradiate" the sample with radiofrequency waves in such a manner as to destroy temporarily the nuclear magnetizations of all except the selected element. The signal-generating pulse was then applied, and a signal was thus received from the single element alone. In 1976 the new technique was described theoretically in a physics journal together with images of simple biological specimens (such as the leg bone of a turkey or a lupine stalk) (Mansfield, Maudsley, and Baines 1976).

Rooted neither in day-to-day professional experience nor in the exigencies of a medical career, medical goals are not "givens" for physicists in the way that, broadly speaking, they are likely to be for physicians. In this case neither physicist was initially committed to any specific medical objective for magnetic resonance imaging. The literature connecting NMR phenomena to carcinogenesis, to be found largely in cancer journals, must have been unfamiliar to them. Each soon encountered this literature, however, and began

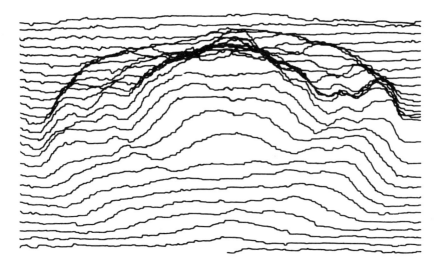

Figure 6.1
An early experiment on the principles of NMR imaging carried out at the
University of Nottingham. This is an image of a cross section of the roots of a
spring onion; the apparent height of the image is proportional to the density of
free water at the corresponding point in the sample. From Hinshaw 1976, 3717.

to conceive of his work in terms of medical applications. Each was
rapidly enthused by the possibility that NMR imaging might prove
to be a remarkable new means of cancer detection. Even though by
1974 Damadian's initial claim of an association between relaxation
time and cancer had become controversial, it nevertheless had
considerable allure. "The spin-lattice relaxation time can be measured as a function of position in the specimen. Such measurements
could be the basis of a valuable diagnostic method for malignant
tumor tissue since the spin-lattice relaxation time is known to be
different from that in normal tissue" (Mansfield, Maudsley, and
Baines 1976).

Mansfield continued to pursue the cancer connection, his work
coming to focus precisely on the problem of tumor detection.
Andrew, by contrast, moved quite rapidly to a view of NMR as a
general imaging modality, and his publications soon offered a less
positive reading of the cancer literature.

In each case, a growing orientation toward unfamiliar medical
objectives began to impose a new discipline, both technical and
social, on the work of the Nottingham physicists. This is associated
with a certain shift in the social locus of the work—less firmly rooted

in physics, we see the constitution of more clearly hybrid structures. For example when, shortly afterward, funds for new larger magnets were required both physicists deployed the "medical promise" of the technology. Though they were physicists, both Andrew/Hinshaw and Mansfield applied in 1975 (successfully) to the Medical Research Council for funds. We see this new discipline also in the changing problematizations characterizing the research. It became necessary not only to build up new contacts with the medical world but also to conceive of specimens both larger and more complex, and to devote new attention to imaging speed. The commercial laboratory instruments being used could accommodate specimens no larger than 1 cm. "It was quite clear," Professor Andrew explained "that if it was going to be useful it would have to be applied to objects bigger than 1 cm." The way in which this scaling up was considered nevertheless continued to show the marks of its roots in physics. We see the attempt at each stage scientifically to prepare the ground for the next. What new problems were to be expected with a larger apparatus? We see concern with the adequacy of the imaging method and a studied awareness of the various possibilities. These reflections could still be turned into publishable physics.

When Andrew and Hinshaw received their MRC funds, they set about constructing an apparatus capable of accommodating larger samples: up to 10 cm diameter. They soon found that with increased size more time was needed to produce an image. Here too we see the growing importance of medical considerations, for it is within the clinical context that imaging time becomes potentially problematic, its reduction desirable. This led Hinshaw and Andrew to seek a faster scanning method. They came up with a variant of Hinshaw's sensitive-point method, using two intersecting coils to define a "sensitive line." The image could be built up line-by-line rather than point-by-point. Replacing the tracer pen of figure 6.1 by an oscilloscope, images at this stage could typically be obtained in two minutes.

Gradually samples studied were becoming not only larger but also morphologically more complex. As it became necessary to provide anatomical drawings or other means of validating NMR images, and as audiences to be addressed changed, medical collaboration became all the more vital. In 1977 Hinshaw et al. published an in vivo image through the wrist in *Nature* and Mansfield and Maudsley published the image of a finger in the *British Journal of Radiology*. (For a slightly later example see figure 6.2.)

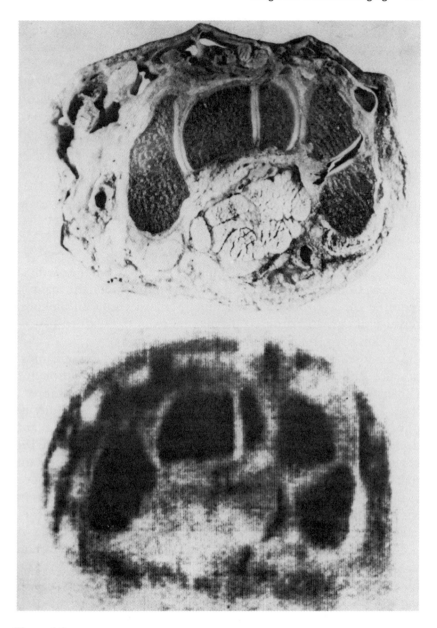

Figure 6.2
An early NMR section through the wrist of Dr. P. Bottomley, with a photograph
of the matching section from a cadaver wrist. From Andrew 1980, plate 2.

For both Mallard and Damadian, by contrast, medical goals were paramount—not only rooted in experience but also sustained by the environments in which they worked. Neither had any interest in the gradual and reflexive scaling-up of the apparatus that occupied Andrew, Hinshaw, and Mansfield. Mallard's immediate reaction when in 1974 a mouse was successfully imaged was, he explained, "We've got to build a whole body machine." There was no question of gradually refining the technique through study of physical phantoms. "I'm not interested in fingers and apples and oranges," Mallard told me, "but in real patients." Nor was the question of what the imaging parameter should be a matter of doubt. Looking back, Mallard stressed his initial conviction, on the basis of that first mouse image, that what they had to capture was not just "an" NMR parameter, not signal strength, but T_1. "T_1 gave us pathology."

When in 1976 he received the funds he needed from the Medical Research Council, Mallard immediately commissioned the magnet required for a whole body machine. Mallard chose an air-core-resistive design and ordered his magnet from the Oxford Instrument Company. It arrived in 1977.

We can see how Mallard's experience with the realities of clinical practice affected his priorities from the outset. Recall how for Hinshaw and Andrew the search for an imaging method different from Lauterbur's had been the challenge to their ingenuity as physicists; this had been the starting point for their work. Mallard decided quite rapidly to modify his imaging method, but for reasons to do with the exigencies of *clinical practice*. It was important that the method not be too sensitive either to patient movement or to field nonuniformities (due to the effect of, for example, iron in the building work). Mallard and Hutchison came up with what they called "selective excitation," a method akin to Mansfield's.

The work both of Damadian and Mallard reflected their common vision of NMR as a means of tumor detection. Damadian and Mallard shared the same concern with the relations between NMR parameters and pathology. But Damadian was necessarily worried by the controversy surrounding his earlier results. Previous work, including his own, had depended upon the characterization of a sample by a single T_1 value. In subsequent work Damadian criticized what he now designated "T_1 null," as "a relatively inaccurate measurement, due to the heterogeneity of the tissue sample" (Koutcher, Goldsmith, and Damadian 1978). A "malignancy index" was defined, in place of a single T value, based on graphical analysis of both T_1 and T_2 readings taken at different signal conditions. In experiments this index was effectively able to discriminate normal from pathological colon, lung, and breast tissue where T_1 null failed. "Regardless of the cause," Damadian and his colleagues emphasized, "the data presented here indicate that NMR is capable of distinguishing normal tissue from malignant pathology with a high degree of consistency."

Though Damadian and Mallard shared this fundamental commitment, there were nevertheless differences in their respective research programs. Damadian was endlessly pressed for funds. In 1974 his fortunes changed—albeit briefly. He received a further small grant from the NIH. In February he was awarded the patent on the method of NMR scanning for which he had applied in 1972. U.S. patent number 3,789,832 was for "Apparatus and Method for Detecting Cancer in Tissue." Recall that Damadian had conceived an imaging technique he called FONAR (Field Focusing Nuclear Magnetic Resonance), which *shaped* both magnetic and radio frequency fields to define a single imaging "window" 1 to 3 mm in diameter. Scanning was accomplished by electronically moving this window, the only point at which the conditions for resonance were satisfied, through the sample. In early 1976, using his FONAR technique, Damadian succeeded in producing an image through the thorax of a live mouse, which he published in *Science* (Damadian et al. 1976). The paper proudly concluded "Having originally introduced the NMR method chiefly for the purpose of noninvasively detecting internal tumors in humans, we have succeeded in obtaining the first images of a tumor in a live animal." Reassured by this success, but without the funds to purchase a large enough magnet, Damadian decided that the whole body scanner he wanted could be produced in one way only. They would have to build it, including the magnet, themselves.

The magnet was going to be crucial, and Damadian and his colleagues had absolutely no experience in building a giant magnet. Damadian decided to build a superconducting magnet because he thought building a resistive magnet (such as the British groups had purchased) would be too difficult, and he was unsure if they could be made to produce images. "I tried to build the most powerful magnet I could dream of building. This could only be possible with superconductivity, so I decided on a five thousand gauss superconducting magnet" (Kleinfield, 119).

"For the better part of a year," Minkoff recounts, "Mike Goldsmith, Dr. Damadian, and I built Indomitable, as we called the first
human scanner" (Minkoff 1985). The odyssey—one of weeks of
winding niobium-titanium wire onto giant aluminum hoops, fifty-
two layers of it, the construction of the giant dewar flasks—ten feet
tall and six feet wide, which would contain liquid helium coolant
(and liquid nitrogen to keep the helium cool)—is described graphi-
cally in Kleinfield's book. In May 1977 their machine was complete:
"a mammoth metal device . . . a big circular thing that vaguely
resembled the slowly turning wheel in a fun house that you try to
walk through without losing your balance. . . . A narrow wooden
plank pierced the center of the wheel. . . . The cardboard coil was
attached to it. The whole thing looked like a mediaeval torture
chamber" (Kleinfield, 167). The FONAR principle now involved
mechanically moving the subject through the resonance point. The
first attempt at a chest image, with Damadian himself as the subject,
was an embarrassing failure, although the subject seemed to suffer
no ill effects. In July, with Minkoff as subject, a successful image was
finally obtained. Imaging his chest required moving him into sixty-
four positions and kept him in Indomitable for four hours and forty
five minutes. Damadian immediately wrote up the experiment,
submitting it to *Physiological Chemistry and Physics* (Damadian 1977).
He deliberately bypassed the more prestigious journals, such as
Nature and *Science*, because he feared any delay would let his rivals
in first. At the same time a news conference announced FONAR to
a broader public.

Although comparison with the exploratory period of ultrasound
seemed to suggest itself by virtue of a similar diversity in constituent
research programs, the cases are very different. Beyond the effects
of twenty-five years of progress in microelectronics is the crucial fact
that in contrast to the ultrasound pioneers, Lauterbur, Andrew,
Mansfield, and other NMR scientists were not medical men. For
Damadian and for Mallard, as for Wild, the possibility of early tumor
detection was of fundamental relevance for their work. In the
ultrasound case the initial contrast, in the exploratory phase, was
with the radiologist Howry. Howry drew his sense of ultrasound as
a general-purpose imaging modality from his experience of the
limitations of radiological techniques. The contrast here is with
Andrew, Hinshaw, and Mansfield who, like Lauterbur, derived no
specific medical commitment from professional experience. Their

sense of the possible medical applications of MRI emerged gradually from reading and from their new medical contacts. They did not provide any alternative context of professional practice within which the technology could take parallel form. In the ultrasound case the notion of replication had little relevance. Initial awareness of each other's work was low among early investigators, and few among those discussed were inspired to attempt to replicate the experiments of others. In the MRI case there were clear attempts at replication of what Damadian did and then what Lauterbur did. But the significant element extracted and accepted or problematized differed from one follower to the other, according to the context within which each was working.

The general reaction to Damadian's success seems to have been mixed. Goldsmith reports: "At first the criticism started to coalesce around the argument that, yes, you can determine the difference between cancerous tissue and normal tissue, but that's not the question. The question is, Can you determine the difference between cancerous tissue and other abnormal pathologies? There was very little data by anyone on this. . . . I think there was a lot of resistance from the pathologists who had a vested interest in the current technology" (Kleinfield, 191).

Damadian had produced the image for which he and others had been striving. But its significance was by no means agreed upon.

NMR Imaging: The Development Phase, 1977–1984

By 1977/78, despite the seemingly slight enthusiasm that greeted Damadian's initial success, clear signs of interest were emerging outside the small group of MRI researchers. The publication of a number of striking images attracted attention, and then growing expectations, within both the medical community and industry. As the implications of possible use in the hospital became clearer, researchers were faced with new problems, new choices. As commercial interests manifested themselves, they exerted a powerful effect on the sociocognitive structures within which research was being pursued.

By virtue of the sums of money committed, we are worlds away from the development phases of ultrasound and thermography. Nor did the commercial dynamic that characterized CT development manifest itself here. By the time the CT development phase began, its principal market, and by the same token principal use, had been fixed. That was in 1973. Neuroradiologists awaited arrival of the equipment with bated breath, while firms supplying them (or hoping to supply them) with imaging equipment worked frantically to get it to market. The beginning of the development phase of MRI in 1977 was different. For one thing, professional and industrial attention was still largely fixed on CT scanning. For another, it was far from clear how nuclear magnetic resonance imaging, were it to prove feasible, was to be integrated into medical practice. If, as some hoped, its application was to be to cancer diagnosis, then in whose province would it fall? Goldsmith refers to the skepticism of pathologists, not of radiologists. What was the market to be, and what the uses? Between 1977 and the early 1980s the claims of radiologists to the new device would come to carry sufficient conviction for them to become its principal market. Even so, a sufficient measure of agreement regarding potential uses for further development work to proceed confidently would be lacking. For the first time market control by radiologists was not accompanied by clear signals regarding the use to which the device was to be put. This had important implications for development work, since both design (for example choice of magnet) and choice of imaging sequence (the kind of image generated) depend upon assumptions regarding future use.[8]

This phase of the career of MRI displays two interrelated themes: first, the responses of the existing academic research programs to growing medical and industrial interest and second, the initiatives taken by industry itself. The locus of research and development work shifts to industry, but gradually, and industrial strategies require complex links with academic research. We saw that coalitions involving medical researchers and industry, often deriving from chance encounters within particular local milieux, were vital to the development phases of both ultrasound and thermography. By the end of the 1970s, with university-industry relations having become a matter for protracted analysis and for initiatives at all levels (National Science Board 1982, Organization for Economic Co-operation and Development 1984), there was little scope for chance encounters of any kind.

For investigators who had focused on cancer diagnosis from the outset, growing interest within the medical community did not entail any immediate change of focus. Clinical uses could be established more precisely by studying the values of NMR parameters for different types of tissue, both normal and diseased. Crucial here were comparisons of adjacent soft tissues: the differences between relaxation values of gray and white human brain tissue, or kidney medulla and cortex values. For Damadian such tissue characterization had of course been his starting point. Mallard initiated a series of studies designed to establish T_1 values for different types of normal tissue, both animal and human. For example, fifty different types of rabbit tissue were examined, and a "T_1 map" of the rabbit was produced (Mallard et al. 1980). Other in vitro studies examined pathological tissue. For example, one such study followed inflammatory and immunological processes in two batches of rats that had foreign tissue implanted in their thighs—normal tissue in one batch , tumor tissue in the other (Ling et al. 1979).

By contrast, for groups that had begun more out of technical curiosity, the realization that clinical application of MRI really was in the offing had major technical and social implications. In the Nottingham work we can see trends begun in the exploratory phase carried further.[9] Collaboration with medical colleagues was becoming more and more vital.

In 1977 neither Andrew and Hinshaw nor Mansfield had reached the stage of whole body imaging. They had however been making rapid progress in improving the quality of their images and were looking at increasingly complex anatomical structures. Here arose the necessity of closer links with those knowledgeable about such structures. As Andrew put it, "as soon as we started doing experiments on limbs we really needed to know what was in there." As interpretations of images of fingers, wrists, and arms came to be presented to medical audiences, accuracy of interpretation became essential. It was all the more essential if, as in the case of Andrew and Hinshaw, magnetic resonance imaging was coming to be seen as a general imaging modality—a technique for radiologists. What was (initially) at issue in presenting such a technique to radiologists was precisely the quality of its anatomical representations.[10]

Having already published a successful image of a human wrist (the quality of which alarmed Damadian), in 1978 Hinshaw succeeded in producing an excellent image of the head of a rabbit (Hinshaw et al. 1978), which he published in the *British Journal of Radiology*. With this paper we can see the clinical—and specifically the radiological—context looming larger. This was apparent not only from the choice of journal but also from the language of the paper. It was no longer simply that publications showed "the close correspondence between the NMR images and the anatomical sections." "For the first time," Hinshaw and Andrew claimed, "images produced by NMR can be compared in quality to those produced by x-ray tomography." The time taken to produce the image of the rabbit's head was eleven to eighteen minutes, depending on the trade-off between imaging time and "both noise and resolution" in a given run. It was the search for improved image quality (without having to make further concessions on imaging time) that was the focus for much of the research. At this point, too, thoughts were turning to scaling up the apparatus from the existing 10 cm working aperture to whole body size (about 60 cm diameter) and to the problems of sample heating and image distortion that this would pose. Despite the fact that with the expiration of the MRC grant two members of his team left for the United States (where they joined industrial MRI development programs), Andrew decided "to go up to whole body size." Funding from the private Wolfson Foundation permitted commissioning the large magnet needed for the whole body machine. Rethinking the choice of magnet shows anticipation of future hospital use, where capital and running costs, and the need for adaptation of buildings, become vital. Andrew and his colleagues decided to abandon the iron-cored electromagnet design of the 10 cm instrument. A magnet of that type, and of the necessary aperture, would weigh some 100,000 kilograms and have enormous power consumption. Considering alternatives, they decided on an air-core-resistive electromagnet that cost some $500,000 less than a superconducting magnet. Uncertainty regarding likely medical application had implications for design, especially of the computer software. It was seen as necessary to allow for changing the kind of image produced, "since it is by no means clear whether detailed NMR data from a point, high resolution in the image, or speed of image production will be the most useful attribute of NMR imaging" (Moore and Holland 1980).

For Mansfield, too, body scanning was approaching. In 1978, using his new 64.5 cm magnet, a "thin cross sectional slice of a live human body" (his own abdomen) was obtained. It took forty minutes of imaging time (Mansfield et al. 1978). Imaging speed was an increasing preoccupation of the Mansfield group, giving the work a rather different thrust than Andrew's work.

Scan time was an important issue for Mallard in a different sense, deriving its significance not from the rhythm of the heart but from the rhythm of the hospital. "Just look at it from the point of view of providing a service," he explained. Preparing the patient for the scan is a time-consuming business, "by the time you've done all this the time in the machine is negligible." Judged in this way, there is little sense in trying to reduce scan times still further. By the same token, the movement of organs (such as the heart) was something to be corrected for (since it spoils the images), rather than something to be pursued for the technical challenge it poses. Thus, in reporting their first whole body images to a 1979 symposium at the Royal Society (figure 6.3), the Aberdeen group pointed to a sort of bar that occurred in the picture: "a movement artefact caused by heartbeats, one of many problems yet to be overcome" (Mallard et al. 1980).

Growing medical interest had not passed unnoticed in industry. Firms with established expertise in the manufacture of laboratory NMR equipment began to see the attractive possibility of a new market. That was not all. By 1977 the excitement of CT scanning had abated, with markets stabilizing. The multinational x-ray companies (including GE, Philips, and Siemens), having slowly but surely taken control of the CT market, began to turn their attention to the new opportunity that magnetic resonance imaging seemed to offer. Even for firms of this size it was going to be an expensive opportunity, with the initial R&D capitalization costs for a firm with no NMR expertise estimated as between $4 million and $10 million. The total costs of developing a system from scratch could be of the order of $10 million to 20 million. Given the high costs, and the uncertainties of a market that did not as yet exist, there was a powerful inducement to try to reduce costs and uncertainties. The cost and uncertainty-reducing strategies of these giant firms radically affected the (hybrid) sociocognitive structures within which the academic groups worked.

For the UK university-based groups, growing industrial interest both deprived them of experienced staff and provided funding for extension of their work. Both at the level of structures and of problematizations it was rather gradual. First there was a movement into the hospital and the accumulation of clinical experience, and then the development of significant (even dominant) structural relations with industry.

In the physics department of Nottingham University, Andrew was getting good results on patients who were brought there from the university hospital. It was becoming necessary to consider the implications of relocating the instrument in a hospital setting.

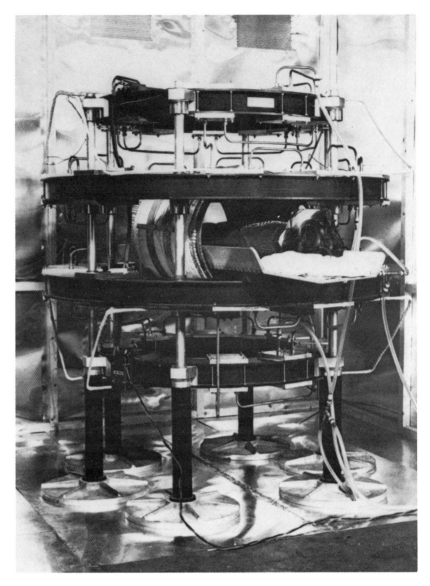

Figure 6.3
Experimental whole body instrument built at the University of Aberdeen
(January 1979). From Mallard et al. 1980, plate 3.

These were still more than simply logistic. Consider the construction of the building in which the device might be housed. Weight was not the only problem. Iron objects in the environment (such as steel in the building's structure) can affect the homogeneity of the magnetic field and hence the functioning of an NMR scanner. If major adaptation of hospital buildings was required, the device could be a financial nonstarter. It would therefore be useful if sensitivity to magnetic field nonuniformity could be as low as possible. This had to be made a priority for development work.

By 1981, results being obtained were so promising that the Andrew group felt that further progress depended on clinical trials being started. On the other hand clinical trials based in the hospital were hardly compatible with further development work, better done with the instrument where it stood in the physics lab. The only way of resolving the dilemma seemed to be construction of a second machine, to be installed for clinical use in the university hospital. And that, of course, would require considerable financial resources. At this point an important connection was made. One of the research group, "casting around," "entered into a relationship with Picker, which was *just* being acquired by GEC. . . . They [have] agreed to fund a new system in our hospital"[11] (Andrew interview).

Something similar was happening in Aberdeen, where Mallard, too, was led to seek industrial funding for a second machine. In the spring of 1980 the Mallard group succeeded in getting its preferred "spin warp method" working to their satisfaction. In August 1980 they obtained "magnificently successful images from our first patient . . . just in time for the big IAEA Radionuclide Imaging Conference at Heidelberg" (Mallard, MSS). By late 1980 work in Aberdeen was beginning to outgrow the instrument, just as was happening with Andrew in Nottingham. Patients were being referred from other hospitals to such an extent that they kept the machine busy three days a week. By the end of 1981 more than 100 patients had been examined, and together with F. W. Smith, a specialist in nuclear medicine, a number of clinical studies (e.g., of the liver) had been published in the *Lancet* and elsewhere.

"We had the biggest amount of clinical experience in the world," Mallard explained, "but we were getting behind technically." It was becoming apparent that a second machine would have to be constructed for clinical use, leaving the first free for further technical development work.

The Aberdeen program, like those in Nottingham, had been funded for much of its life by the Medical Research Council. As is usually the case with work funded at universities by the research councils in Britain, the patent on the instrument developed was taken out by the National Research Development Corporation (now part of the British Technology Group). The NRDC would license the patent, if it could, and there would be a royalty-sharing agreement. Professor Mallard also had some support from the Technicare company in the form of gifts of instrumentation. When, at the end of 1980, Mallard felt that the only way to resolve conflicting demands on the instrument was to construct a second, the question of funding arose once more. The Medical Research Council refused to fund a second machine for clinical use on the grounds that their funds were not for the construction of hospital equipment. The NRDC, which held the patents, was prepared to make a loan. But how was a loan to be repaid? Charging patients, which was one way of recouping development costs in the United States and was proposed by the NRDC, was not acceptable in an NHS hospital. A real possibility was Technicare, then anxious to improve its position in the new technology. For the Aberdeen scientists, however, the company's conditions were not acceptable, for they wished temporarily to relocate the whole research group at their research establishment in the United States.[12]

At this point Professor Mallard went to lecture in Japan where, he explained, the NMR imaging idea had not yet been picked up. Japanese companies were greatly interested, and many were willing to put up the funds Mallard needed. It was agreed eventually to work with Asahi, a large chemical company that saw NMR as a means of entering the diagnostic imaging market. Asahi did not suggest relocating the research in Japan, but would send its own scientists to Aberdeen to learn the technique. Moreover, the company was prepared to let the Aberdeen group collaborate in addition with any British company it chose. By mid 1981 funds had been received, and the MkII instrument was under construction. Though Mallard was very anxious to establish a parallel relationship with a British company, no suitable partner could be found. With approval and support from a variety of (largely public) sources, a new Aberdeen-based company, M&D Technology Ltd., was established to manufacture the Aberdeen-designed instrument.

Damadian reacted much more rapidly and quite differently than the British groups. He saw growing industrial interest as a new and threatening form of competition. By the end of 1977 he was sure not only that a market for NMR scanners was emerging but also that industry had come to the same conclusion. Contacts with GE and Johnson & Johnson (which seems to have suggested that Damadian might become head of a new subsidiary) proved abortive (Kleinfield, 199–200). Concerned that big business would steal his idea, and anxious to leave the university, Damadian decided to set up his own company. In March 1978, adopting the name FONAR Corporation for the new company, he set about raising funds by selling stock.[13] The move to commercial operations led to an important reconceptualization in Damadian's FONAR design. Like Andrew, Damadian reconsidered his choice of magnet. Hitherto he had worked with superconducting magnets. Thinking about making instruments for sale, he decided to build a permanent magnet. No one had ever built a permanent magnet as large as the one the FONAR instrument would need, and many advised Damadian against changing his design. Yet, he was convinced that superconducting magnets would be too expensive (given their great appetite for liquid helium and liquid nitrogen) and too troublesome for hospitals to bother with. By December 1978 he had produced a 500-gauss permanent magnet.

Transformation of these various prototypes into commercial devices was taking place through a variety of processes. If we continue to think in terms of the hybrid sociocognitive structures within which research was going on, it is clear that they had changed almost beyond recognition, increasingly dominated by industrial actors and norms. Both Damadian and Mallard had set up companies to manufacture devices according to their designs. The Aberdeen technology was also the means for Asahi to enter the diagnostic imaging market. Other companies were seeking ways of reducing the barriers to entry by obtaining access to existing expertise located in the universities. This is what Technicare did in appointing Hinshaw to run their development program. Similarly GE appointed Paul Bottomley, another ex-member of Andrew's group, to a senior position in its development program. Diasonics, a major manufacturer of computerized ultrasound equipment, took a share in an R&D program based in the radiology department at the University of California at San Francisco when Pfizer, which had been supporting it, decided to pull out. Another way of reducing the financial barriers to entry was to share costs. This was the strategy adopted by EMI.

EMI was one of the first companies to respond to the possibilities of NMR. In 1976, still flush with its CT scanner success, EMI began development of a magnetic resonance scanner in its central research laboratory. Work was under the direction of Dr. Hugh Clow, an ex-Nottingham physicist who specialized in magnetic materials, and Dr. Ian Young, head of the data systems department. On the basis of a preliminary theoretical investigation, the EMI group opted first for a resistive magnet and for Lauterbur's reconstruction method. Progress was rapid, and it was this laboratory—focusing on the skull just as they had so successfully done with CT scanning— that first succeeded in publishing an NMR brain scan. Under the heading "Britain's brains produce first NMR scans," this scan of Dr. Clow's skull appeared in the weekly *New Scientist* in November 1978. On the basis of this success, EMI approached the DHSS for support of a number of prototypes for clinical evaluation, following the same strategy as with the CT head scanner. It was agreed to share the cost of one model, to be installed at London's famous Hammersmith Hospital. In designing it, the EMI scientists also opted for major changes: to build a whole body machine based on a superconducting magnet.

Damadian produced his first model for sale, and the FONAR Corporation revealed itself to the world in March 1980. Called the QED 80, this was the first commercial magnetic resonance scanner, with the appearance (as Kleinfield puts it) of a "sleek mobile home that had had a hole gnawed out of the center of it." In April 1980 a prototype of the QED 80 was taken to the annual meeting of the American Roentgen Ray Society. By early 1981 four prototypes had been placed for evaluation: in Cleveland, Ohio; in Japan; and in the Universities of Milan and Mexico (Kleinfield, 212).

The x-ray industry, mobilizing its vast scientific and financial resources and its structural links with academic radiology, was catching up fast. Philips had learned of the work of Lauterbur and Damadian in 1976/77. Lauterbur visited the firm, and a few small-scale experiments were successfully carried out.[14] In early 1979, on the basis of a short technical and financial study, Philips assembled a group of four or five scientists with the task of designing and constructing a whole body system as quickly and cheaply as possible. By mid 1981 Philips had a prototype that gave reasonable whole body images. From physical phantoms the experimental focus moved to human volunteers and then to patients. In collaboration with nearby medical schools some hundreds of cases, mainly neurological, were scanned with this first prototype located at Philips Medical Systems laboratory.

The EMI system, named Neptune, was installed in January 1981. Its future was to be determined less by its performance than by corporate strategy. On the breakup of EMI Medical the new company (Thorn-EMI) proposed to sell the NMR program to GE (which is of course American). For some time work at Hammersmith Hospital continued under considerable uncertainty. The sale to GE was blocked by the British government, at which point Picker International appeared as an alternative purchaser. Picker International already had experience with resistive magnet systems, having both developed its own and supported Andrew's in Nottingham. Why should they have taken over EMI's program of work on a superconducting system? The rationale again has to do with the uncertainties that continued to surround NMR imaging technology.

By 1981 industry was confident enough of the existence of a market, if only for research instruments, large enough to justify the necessary investment. But there was still uncertainty regarding trade-offs in design; views differed, for example, as to whether the extra costs of the superconducting magnet were justified by the higher field strengths realizable. There was also uncertainty regarding likely uses. Each magnet system had its particular advantages and disadvantages in terms of capital costs, running costs, the particular uses to which the system would be put, and so on (American Hospital Association 1983). Large firms with sufficient resources (including GE, Philips, and Technicare) hedged their bets. Picker International, too, had experience with the two major technological variants.

By 1983, a survey by the Office of Technology Assessment showed, there were some twenty firms engaged in NMR imager development programs, of which six (including FONAR, M&D Technology, and Bruker) were specialized in NMR. Most of the rest were the major x-ray suppliers (including Hitachi, Shimadzu, and Toshiba). Of twelve firms prepared to say, four (which had recently entered the market) had spent less than $5 million on R&D , whereas five firms had spent in excess of $20 million (OTA 1983). The market projections seemed to justify this level of investment. A 1981 assessment of the market in the United States was for the placement of 15 instruments in 1983 (giving sales of $4 million at 1979 prices), rising to 550 instruments ($105 million at 1979 prices) in 1988 (Hamilton 1982). A slightly later forecast was for 15 instruments worldwide in 1982 (which proved accurate), rising to 1250 units in 1988. The air of expectation was spreading well beyond the professional community. By 1983 even the popular press was reporting on the new technique with captions such as "Looking beyond the X-ray" (London *Sunday Times*, June 1983) or "A Diagnostic Dream Machine" (*Boston Globe*, April 1983).

The rules of the game were becoming those of multibillion dollar industry, and for some of the first researchers in the field they were no longer agreeable. Professor Andrew, for example, began to feel it was time to quit and go back to physics. For Damadian, on the other hand, the pace was intensifying. It was becoming more and more essential to look at FONAR's performance in terms of what *competing* systems had to offer. Kleinfield describes the feeling of panic Diasonic's images had aroused in Damadian, who "had no idea such fabulous images were possible" (p. 214). That had been in 1981. The only possibility, Damadian realized, was to move to much higher magnet strengths and thereby improve the quality of his own pictures. But to produce a permanent magnet yielding a 3000-gauss field (thus comparable with Diasonics' superconducting magnet), *and* within a year (before FONAR was written off in the industry), was quite a task. Weight would have to be discounted, for it would weigh 100 tons, ten times the weight of the existing 500-gauss magnet. Somehow, it seems, Damadian managed it, and his Beta 3000 model was well received at the important Radiological Society meeting of 1982.

A major market in "investigational" devices had been established. Since there was as yet no FDA approval, the magnetic resonance imagers could not be sold in the United States for routine patient care, and hospitals could not be reimbursed for scanning patients. But they could be sold for clinical investigation. In terms of the definitions being used in this book, we are still in the development phase. However we now have the new notion of a clearly recognized market for investigational devices. If we compare this technology, in the early 1980s, with those that have gone before, then commercial competition was clearly becoming an increasingly important consideration within the development phase. For many firms it seems to have been an assessment of the potential of this investigational market that justified the commitment of major resources. In 1983, thirty-nine machines were sold worldwide, twenty-eight of them in the United States (Kleinfield, 225).

FONAR was the only company making a system based on a permanent magnet. Convinced that his system would ultimately prove the more reliable, Damadian attributed hospitals' seeming preference for competing systems to "corporate power" (Kleinfield, 228). Doctors could not accept that he could be right where GE, Technicare, Siemens and the rest were wrong, he is quoted as saying. The reality confronting Damadian was of course the assertion of its collective interest by the interorganizational structure centered on x-ray technology.

In 1984 the FDA granted market approval to FONAR, Diasonics, Picker International, and Technicare for their NMR systems, and in 1985 to GE.

Assessment and Control of MRI since 1981: Quis Custodiat Custodes Ipsos?

Earlier chapters have confirmed how bound up with clinical context and professional control are the questions posed in the evaluation phase of a new imaging technology. We saw that the CT scanner, designed from the start to function within a specific practice, was in this sense unproblematic. There was little or no disagreement either about the kinds of questions to be addressed to the new technique—for example, what *precisely* does the CT scanner do, and how well does it do it—or about the terms or the methods with which answers were to be established. Ultrasound was very much more complex, by virtue of its parallel constitution within a variety of distinct medical practices, each of which gave rise to its own agenda for clinical research and assessment. Because of this, possible conflict between medical specialties over its assessment and control was averted.

The MRI situation has some novel elements. When Damadian first suggested a nuclear magnetic resonance scanner for medical purposes in 1971, he had seen it as an important addition to the hospital pathology laboratory. In 1977, when the first whole body FONAR image was published, it was the unenthusiastic response of the pathologists that so chagrined Goldsmith. Yet when Damadian put the QED 80 on the market in 1980, it was the radiologists to whom he demonstrated it. Radiology seemed to offer an established market, and its practitioners were enthusiastic. For them, nuclear magnetic resonance was becoming, in the words of a leading article in *Radiology*, "Another new frontier" (Evens 1980). Despite this gradual assertion of radiological interests, the development phase was marked by persistent uncertainty as to the uses to which the technique would be put. This uncertainty had implications both for hardware design and for the method, or sequence, used in generating the image. As we begin to look at assessment of the technique, at the bases on which its utility was being assessed, it becomes clear that something other than uncertainty was also involved.

In an important lecture to the Royal College of Radiology in 1981 John Mallard told his audience:

The human body is extremely complex. When, in addition, we first attempt to image a new property such as proton magnetic resonance, there is bound to be difficulty in interpreting the results. To these difficulties, further confusion may be added if early NMR images are compared with other images derived from completely different tissue properties, e.g. X-ray absorption, radiopharmaceutical concentration, or ultrasound characteristic impedance. It has always been our intention, confirmed by the early animal images . . . to attempt to image T_1 separately from proton concentration. With this in mind, we have carried out a biological back-up program of T_1 measurement on normal and pathological tissues to ease the problem of image interpretation and to find pointers toward the most fruitful fields for the application of NMR imaging. (Mallard 1981)

Mallard was arguing that the interpretation of images based on a new property (such as NMR) is not straightforward and cannot automatically be based on comparison with images derived from other means. This has implications for establishment of the principal uses of the technique. Above all, Mallard was claiming that it must be through *biological research* that both uses and interpretation of images had to be pursued. These were the functions of the "biological back-up program" being carried out in Aberdeen. The message to radiologists was clear: You have no authority in the interpretation of NMR images, which depends upon biological research in which we, not you, are expert.

Initial marketing of the technology was nevertheless to radiologists, whose enthusiasm reflects an important interest in rapidly establishing mastery of the technology. Unlike practitioners of other specialties, who might gradually come to recognize its possibilities, radiologists have no reason to wait upon the results of evaluations. To the contrary: we have seen their reputational interest in being among the first to publish studies making use of just such a new device. Most of the early clinical assessments of the technique appeared in the radiological literature. For the American Hospital Association, this was "because of its anatomical imaging abilities," even though "it also offers exceptional biochemical and metabolic potential." The AHA report goes on, "Some believe that because NMR is more than just an imaging device it requires closer cooperation between spectroscopists, pathologists, biochemists and other physicians outside the radiology field. . . . Some believe nuclear medicine physicians and internists may be best equipped to understand NMR because of their experience in metabolic physiology" (American Hospital Association 1983). Some have jokingly suggested that NMR really stands for "No More Radiologists." Less flippantly, the AHA went on to warn hospital administrators that they must in any case be prepared to defuse tensions that may arise over control of magnetic resonance imaging.

We have encountered relatively little overt *conflict* over the control of a technology. The AHA refers to conflict arising in the day-to-day functioning of an individual institution. In chapter 2, I argued that conflict over an emergent technology that seemed to present a real threat to existing interests could leave its mark on that technology's career. Certainly in the case of MRI considerable interests, both monetary and reputational, are at stake.[15] We expect competition in the marketplace, but how else might such conflict manifest itself? Professor Mallard's warning to radiologists attests to the plausibility of the claim that struggle for control is reflected in competing problematizations. If conflict over an imaging technique becomes manifest, we might therefore expect public disagreement over the questions to be posed in assessing it, and over the methods by which answers are to be sought.

We can see how a gradual sense of the possibilities of MRI in fact began to be established by looking at an early report of the first year's experience with the EMI Neptune. Thirty volunteers and fifty patients had been studied at the Hammersmith Hospital (Young et al. 1982). Three sorts of imaging sequences had been used, producing three kinds of images: one based principally on proton density (the so-called free induction decay, or FID sequence), one based principally on T_1 (the inversion-recovery sequence), and one based principally on T_2 (the spin-echo sequence). These sequences, which can be altered by the computer, produce rather different images. This, of course, is a unique characteristic of MRI, with important implications. "The variety of sequences possible with NMR allows a choice of image to reflect varying degrees of proton density and blood flow, as well as tissue spin-lattice and spin-spin relaxation times. While this gives NMR considerable versatility, much more work will be required to establish the optimum technique for particular conditions and circumstances" (Young et al. 1982).

MRI does not image a conventionally unproblematic "structure." The image can be made to reflect one of a number of subtle, essentially chemical, properties of bodily tissue. As Dr. Young and his radiological colleagues worked through their material, attempting to establish possible clinical uses of the technique *specifically by probing its advantages over CT scanning*, not only the particular organ had to be considered but at the same the particular property and corresponding sequence used to form the image.

Most evaluation studies have been along the lines of this early example. In Britain the Department of Health and Social Security and the Medical Research Council together funded an evaluation program, which included both the Picker models (resistive and superconducting) as well as the Aberdeen M&D Technology instrument. The studies largely focused on comparison of NMR with CT scanning (Hutton, n.d.); that is, the evaluation assumed that assessment of MRI is to be in terms of its place in radiological practice.

Recently, although no other specialty has mounted a direct campaign to wrest control of MRI from the radiologists, open controversy over the manner in which it is being incorporated into their practice has broken out. Recall how radiologists had been scandalized at an early suggestion in the *Lancet* that they were not necessarily the best people to advise on the deployment of CT scanning. The controversy that recently flared up around the assessment of MRI is reminiscent of the issue the *Lancet* editorial and, somewhat later, Creditor and Garrett raised with regard to CT scanning. That it has been much more heated reflects the greater uncertainty of the situation and hence the greater risks involved.

A June 1988 paper in the *Journal of the American Medical Association* from the Technology Assessment Group of the Harvard School of Public Health argued that "Sound and scientifically rigorous evaluation of new technologies is necessary to ensure quality of care and cost-effective use of resources. The indications for the use of MRI must be established as rapidly as possible to avoid the ordering of unnecessary studies and the installation of unneeded MRI units" (Cooper et al. 1988). These authors looked at fifty-four English-language articles published between January 1980 and July 1984 (including the 1982 Neptune paper) that presented results from at least ten patients examined with an MRI scanner. Thirty-four (63 percent) of the articles had appeared in two journals, *Radiology* and the *American Journal of Roentgenology*, and many show the superiority of the technique to others available, especially CT scanning. Cooper et al. assessed the fifty-four studies on the basis of a number of criteria that, they claimed, a scientifically adequate evaluation must meet (randomizing order of tests, measurement of interobserver variability, appropriate presentation of data, and proper use of terms like sensitivity and specificity). All articles scored poorly, with no improvement over the four and a half years examined. "Although none of us are radiologists," the authors of this article wrote, "the methodology of clinical research is universal and applicable as much to radiologic as to other diagnostic procedures." The authors expressed concern over the way in which radiologists seem to establish the utility of such an expensive device: "We are concerned that in the absence of such assessments, a new imaging modality that provides better pictures tends to become the diagnostic modality of choice without objective data to support its use."

The criticism cuts deep. Can radiologists be trusted to assess new devices in a way which the medical world as a whole can accept as legitimate? What sorts of proof should be required of those who look to add to medicine's technological armamentarium, of those consolidating (or exacerbating) medicine's utilization of (or dependence on) modern technology? Is the demand that the contribution of an innovation to some common health goal be established before commitments are made a reasonable one? Or is it now accepted that evaluation properly focus on the search for the most appropriate among the various possible uses? Fundamental commitment to (the benefits of) technological innovation implies resistance to the first of these alternatives. The reaction to Cooper et al., both in *JAMA* and the *American Journal of Roentgenology*, was fast and furious. In a joint letter the editors of the two main radiology journals in the United States (*American Journal of Roentgenology* and *Radiology*) set out a number of lines of defense that were to be deployed in various ways by many other radiologists: the criteria used by Cooper et al. were not realistic; assessment of new technologies is beset by so many difficulties and constraints (lack of resources, unwillingness of patients and physicians to cooperate, hospital administrators unwilling to subsidize research by allowing duplicate examinations, etc.); the purpose of those early studies has been misunderstood. "Most were preliminary works whose purpose was to assess the feasibility of examining patients, to determine optimal techniques, to describe the range of appearances of normal structures, and to develop diagnostic criteria for the presence or absence of disease. They were pathfinding articles meant to point out promising directions" (Berk and Siegelman 1988).

Most of the other responses argued along the same lines: that Cooper et al. had misrepresented the purpose of the studies, that the constraints were extreme, and that resources for proper assessments were scarcely available. Some writers preferred to counterattack, with arguments to the effect that if a technology saves lives and money, it is improper to sit waiting: "Although medical economists and social welfare theoreticians might wish it otherwise, the reality is that subjective and descriptive observations form the basis for the initial evaluation and acceptance of new techniques in imaging and other fields" (Friedman 1988).

Whatever the merits of Cooper et al.'s criteria, whatever the merits or inadequacies of their study, its effects were simple but, in principle, deadly. It drove a wedge between the real process of decision making and its rationalization in the literature. The response of Dr. Berk and Dr. Siegelman was perfectly rational *given* the radiologists' starting point—that it is preferable to focus on finding the uses to which a technology can be put. No one can be sure, however, that this starting point finds general acceptance, that, in these days of cost consciousness, there might not be an obligation to prove that an expensive new technique has some absolute value. Hence the need to present assessment studies as though informed by an alternative set of underlying assumptions.

Feedback: The Further Development of MRI Technology

The career of MRI differs from the careers of other technologies discussed in this book partly by virtue of the radically different climate in which it was brought into being. This is partly to do with U.S. legislation, dating from the mid 1970s, which formally demarcates the experimental from the clinical periods through the FDA approval process. At the same time the stakes had risen beyond anything the scientists and engineers of the 1960s would ever have imagined. Costs of developing MRI, and the entry barriers faced by firms lacking expertise in nuclear magnetic resonance technology, were so great as to force new approaches to assimilating existing expertise. University-based work on the technology was affected both by the specific interests of large corporations and a climate in which university-industry relations acquired new saliency. Other unique features of the constitution of MRI have to do with the nature of the technology. As the existence of distinctive imaging sequences suggests, the technology has properties radically different from those of other technologies discussed. It is this that gives specific character to the feedback phase and at the same time makes difficult any clear delineation of this phase.

Pioneers of both ultrasound and CT scanning based exploratory studies of the possibilities of image formation on work with simple physical phantoms. Such simple phantoms, producing easily comprehensible images, clearly show the structural basis of the image (see figure 3.4). The MRI image does not refer to any unambiguous structural property. Hinshaw and Andrew hinted at this in an early paper. "Studies made using simple phantoms made up of water, glass and air are not very helpful and may even be misleading," they wrote, "since real biological tissue contains a wide range of relaxation times." It is because of the distinctive nature of the MRI imaging principle that the feedback phase, unlike those encountered hitherto, does not focus either on extending applicability of the technique or on increasing speed or resolution.

Some producers, and some researchers, have for a number of years been contemplating the future possibility of nuclear magnetic resonance imaging based on phosphorus or other biologically important elements. Because their NMR signals are very much weaker than those of the proton (^1H), this would be a difficult task requiring, among other things, very high magnetic field strengths. On the other hand phosphorus and sodium compounds are of far greater relevance for current medical priorities than are water, fats, or other hydrogen-rich compounds. From the mid seventies work at Oxford University, Johns Hopkins, and elsewhere had shown NMR analysis of the phosphorus in living tissue to be a valuable means of investigating degenerative changes in the body. A number of phosphorus compounds and complexes play vitally important roles in the functioning of the body, and biochemists had come to use high-resolution NMR spectroscopy (thus not imaging) as a means of following degenerative processes (including stroke and ischemia) (Marx 1978). It is this increasingly important field of interest that underpins a significant comment made in 1979 at a Royal Society Symposium. In a general comment on the imaging papers presented by Professor D. R. Wilkie, an eminent biochemist, asked whether "we could gain enough sensitivity to work not only with water protons, whose significance is largely empirical, but also with more interesting nuclei, such as ^{13}C and ^{31}P." At the same symposium Lauterbur referred to work he and his colleagues had already begun, starting from the same assumption—that "the direct observation of metabolically and structurally significant individual molecular species in n.m.r. zeugmatographic images would be the most desirable approach to medical diagnostic applications" (Lauterbur 1980). Lauterbur had been working with phantoms of a different sort, in which different biologically important phosphorus compounds were contained in the separate compartments of a vessel. Both desirability and route were clear enough from the start. Their work at the end of the 1970s led Lauterbur and his colleagues to conclude that although in vivo phosphorus imaging of humans was as yet technically impossible, nevertheless imaging experiments could at least "assist in the interpretation of the results of ^{31}P NMR studies of heterogeneous objects, such as hearts or kidneys." From the mid 1980s very high field magnets used experimentally (as for example the 15-kilogauss system developed by Philips and Intermagnetics and installed at New York's Columbia Presbyterian Hospital) began to offer wider promise of phosphorus or sodium imaging.[16]

In a short time MRI has become a remarkably active field of research, with its own international societies and journals, and an important industrial presence.[17] There is a hope in some quarters that MRI will become as familiar and as common as the roentgen apparatus now is. The associate editors and editorial advisors listed in a recent number of the journal *Magnetic Resonance Imaging* include scientists from many renowned academic institutions alongside others from General Electric, Picker International, and Philips and Ciba-Geigy. Lauterbur is now a professor of diagnostic imaging. It is far from inconceivable that, together with the PET scanner which held Mr. Sochurek spellbound in chapter 1, MRI will be associated with fundamental change in the organizational structure that for fifty years has been constituted around x-ray technology.

7

Image of Truth Newborn

Introduction

Diagnostic imaging today, with its "uncanny new eyes" deploying some of the most esoteric findings of modern science, seems worlds away from the Roentgen enthusiasm of a century ago. A vast expansion in the scale of medical care has taken place, accompanied by growing specialization, concentration in giant hospitals, and soaring costs. Professional radiologists have become an integral part of this modern clinical practice, their level of activity among the fastest growing of all specialties in the hospital. The industry supplying radiological equipment has been transformed out of all recognition, with production also concentrated in the hands of just a few firms. The Roentgen tube of 1900 was made by the skilled craftsmen of a small local firm. The complex scanners of today are manufactured worldwide by a few giant multinationals. Yet these very obvious differences must not be allowed to conceal certain important underlying continuities. Out of those early enthusiasms grew a practice, and an industry, that became increasingly interdependent as the years went by. Processes of innovation, the search for increasingly subtle and sophisticated ways of viewing the secret spaces of the human body, are intimately connected both with that practice and with that industry, even today. It is those intimate connections that provided the starting point for this book.

My objective has been to try to come to grips with the processes by which Mr. Sochurek's "uncanny new eyes" took shape. Why is the history of diagnostic imaging one of continuous technological innovation? How can we understand the process through which the vision of a new medical image becomes constituted as a part of medical practice? Can we, in fact, speak of *the* process, or is there more than one? What, in particular, is the influence of those established professional and industrial interests that trace their origins back to Wilhelm Roentgen's discovery of 1895?

These questions sustain the view that systematic social scientific inquiry has scarcely done justice to the complexities of modern technology. As I argued at the beginning of the book, technology-in-society all too obviously transcends the theoretical modes within which we typically try to understand it. Organized technological innovation has become one of the most powerful forces of change in a society that sustains it to the hilt, both economically and ideologically. And if the theoretical modes with which social scientists try to understand technological change are crude, those employed by politicians and planners are cruder still by far. Because we do not understand the processes by which technological visions take shape, we have little sense of what the real consequences of backing an innovation project might be.

In order to try to deal with the specific issues raised by innovation in diagnostic imaging, I proposed a scheme in chapter 2 based on three concepts. One of these, that of career, was most obviously of expository value, while the other two (interorganizational structure and problematizations) were intended to point us toward the issues fundamentally involved in the constitution of new imaging technologies. In this final analysis I shall first return to the model of chapter 2. The subsequent part of the chapter sets out what I believe we have learned of the processes by which imaging technologies are constituted.

Fundamentals of the Analysis

The most important thing about the technologies discussed in this book is not the differences between them but what they have in common—the fact of their being imaging technologies. This makes comparison meaningful. This provides the basis for comparison with other sorts of technologies.

I argued that DiMaggio and Powell's notion of the organizational field applied very well to the hospital system and its suppliers, regulators, and clients. Innovation in medical technology is sustained by widespread consensus regarding its importance. This is just as true of drugs and the other equipment deployed in modern hospital care as it is of diagnostic imaging instrumentation. The basis for comparison with other socioeconomic sectors then derives from characterization of the organizations and interorganizational relations making up this field. We would focus, for example, on the nature and relative extent of organizations' dependence on innovation and on modes of integration and interdependence (including for example professional mobility).

Our focus here is on a lower level of organization, on the set of practices and relationships centered on the technology of medical x-ray. By looking at how this structure of (interorganizational) relations came into being, we learned something of its characteristics. Between the end of World War I and the 1930s the interests of an increasingly oligopolistic industry and those of an increasingly professionalized radiology became more and more attuned to each other. There was a common commitment to innovation, embodied in a continuous process of adaptation and specialization in the basic technology. This meant that compared to the early years, innovation came deliberately to reflect the interests of the professionalizing customers. The result was decades of continuous and essentially incremental innovation: faster films, lower voltage tubes, and specialized applications and equipment.

The technology they control provides the basis for both the professional practices and the status of radiologists. Innovation, the extension of their technological armamentarium, is of crucial importance for a variety of reasons. One is financial. We saw how profitable new diagnostic procedures could be. Another is status enhancement—of the profession vis-à-vis other professions and of the hospital vis-à-vis other hospitals. Yet another is individual reputation within the community of medical science. Being one of the first to have access to a new device provides the opportunity to be among the first to publish on its clinical use. Radiologists rely on their manufacturers to supply them with this new technology, and they participate actively in its development. Though the modern x-ray industry is dominated by multinationals with many other lines of business (GE, Philips, Siemens, Toshiba), it is a profitable and substantial area of activity. Since competition takes place on the basis of performance, not price, manufacturers rely on their professional clients for clinical reports attesting to the performance of their particular products. Today, with development costs coming to exceed the resources of even the multinationals, and with governments trying determinedly to hold hospital costs down, the stakes are all the greater as far as any manufacturer is concerned.

The same holds for the radiologists. For decades there was little to worry about. A few large suppliers gained control of the markets, while radiologists' status, resting as it did on control of sophisticated apparatus and unique ability in interpretation, remained unchallenged. As radically new technical possibilities emerged, so too did the possibility of loss of control. New diagnostic techniques based on radically different principles could threaten both industrial oligopoly and professional monopoly. Maintaining the integrity of existing arrangements requires the attempt to impose control simultaneously over manufacture and deployment of a new technology.

The notion of career was introduced as an expository device, as a way of establishing some sort of a chronology with the aid of which the case studies could be set out. The phases defined in chapter 2 delineated sets of activities that in terms both of content (the nature of the work) and locus (the location of the work) had something in common. In that sense the notion was no different from economists' notion of the product cycle though, as I pointed out, career as used here was intended to refer to circumstances both of production and of use. The concept of interorganizational structure provides some rough markers with which we can flesh out our notion of career. Very roughly speaking, it defines two careers, two paths that the vision of a new imaging technology can take. One is that which responds to existing interests and sees the technology coming to reflect existing structures of production and practice. The other is the converse, involving a challenge to existing interests. We shall see that matters are more complex than this. Nevertheless this simplification makes clear how the act of defining the interorganizational structure makes of career something more than an arbitrary chronology. How does it happen that a technological career takes one or the other (or some third) route? What consequences does career have for the form the technology eventually takes? To find out, we need to look carefully at what is going on in each phase.

It took twenty-eight years to bring ultrasound to market (starting from the origins of the exploratory phase in 1937), ten years for thermography (starting from 1955), fourteen years for CT scanning (starting from 1961), and eleven years for MRI (starting from 1973). In part these differences reflect historical change and contingency. The advent of microelectronics meant that certain problems could be solved much more easily in later years. Those working on ultrasound at the beginning of the 1960s were at one time aware of major changes in the offing that could render any device based on older electrical circuitry immediately obsolete. They felt they had to wait. Stimulating technological innovation, including in the medical area, became a priority of governments. Government played a crucial role in organizing the coalition of interests within which EMI first developed the CT scanner, very different from earlier chance encounters. When the idea of nuclear magnetic resonance imaging arose, many of the firms that might have shown immediate interest were still preoccupied with CT scanning, in which they had made major investments not so long before. Willingness to investigate the possibility of the new technique was delayed. However, when we look carefully at what was going on in the various phases of each technological career, there are many differences attributable neither to historical process nor to contingency. This career-in-the-making should show how it is that one technology (like CT scanning) is incorporated into radiological practice while another (ultrasound for example) is not.

The notion of career implies temporality, an unfolding through time, as well as commonality. Innovators' interest in the success of their innovations requires commitment to moving things along from one phase to the next, with each phase providing its own particular goals (working prototype, production process). Successfully steering an imaging device through its accommodation/ diffusion phase amounts to achieving a situation in which the new device is widely and routinely used in diagnostic practice. In economic parlance, success depends on capturing an adequate share of the market. In terms of the norms and values of the professional community, it entails achieving a broad consensus regarding taken-for-grantedness. Ideally the physician's black box must be so trustworthy that applied to the individual patient in routine practice its working can be taken for granted. At the same time it must be so authoritative that used in research it can serve to authenticate, to legitimate, other procedures or devices. It is in the innovators' interest that the diagnostic instrument attain this status—that questions regarding the nature of its performance be set aside and the attention of the medical community be focused on specifying its precise place in practice.[1]

The third central concept, that of problematizations, reflects most generally the need for a means of referring to the analytical detail of career—the level at which we can look to see what is going on. More precisely, I have been using it to refer to the (succession of) questions being asked about the technology within the organizational field in which it is deployed. What is problematic about the technology, and when, for whom, and with what result?

The conventional way of looking at the place of research in the innovation process is to view scientists as technical problem solvers (see for example Gibbons and Johnston 1974). Starting from the notion that it is limitations in performance, experienced in use, that stimulate much incremental innovation, this then becomes the source of problem formulations such as, "How can we improve the speed, or the sensitivity, or whatever of the device in question?" In fact matters are much more complex. Most important from a sociological point of view, the explanatory notions of problems in use or limitations of performance are not solid enough to bear the weight placed upon them. Many of those involved in the operation of a technology will see no problem, or the problems they see will differ, depending upon whether they are medical physicists, radiologists, radiographers, nurses, or hospital administrators. So whose experiences "really count," and how—and for whom—do they become (research) problems?

For example, we know that dissatisfaction with available techniques of radiological investigation of the brain played an important part in the origins of the CT scanner. There is nothing surprising about Oldendorff's or Ambrose's concern with the inadequacies of neuroradiological techniques. Hounsfield—whose role was of course crucial—is a different matter. How did an electronics engineer working for the firm that was recording the Beatles come to share this same concern? There is a similar incongruity in Lauterbur, a chemist, or Andrew, a physicist, trying to develop ways of imaging the human body. How do experts in stereochemistry or magnetism come to have anything to do with sick people? How does the state-of-the-art in infrared detection become a matter of significance for a cancer surgeon like Lawson or Lloyd Williams? To get at riddles like these, we must move beyond the bare notion that certain problems arose and were solved. To understand the dynamics of the technologies, we must also understand *how and for whom* their possibilities, and their functioning, became problems.

As we look in detail at the research through which the technology was given shape, at the priorities and the choices made, we see career unfolding. We see models giving way to prototypes and laboratory research giving way to clinical research. We see scientific publications claiming priority for what has been done, and we may see other scientists attempting to produce the same effect by replicating the work. The questions being posed in research change from "can I get a picture," for example, to "can I get a picture of good enough quality," to "what will the instrument actually do?" A principal claim of this book is that there are patterns in these evolving problematizations and these patterns are associated with (alternative) careers. It is in these patterns, in the assumptions on which sequences of problematizations rest, the resources brought to bear in tackling them, that we see the unfolding of career.

Central to this perspective is thus the idea that alternative sequences of problematizations are associated with alternative career structures. To see how this might work, consider the notion of replication. At various points I have compared attempts at replicating a new experimentally produced procedure or effect with Collins's discussion of the TEA laser (Collins 1974). Collins looked in detail at a set of experimental physicists in different laboratories all engaged in "the same problem": in trying to build a gas laser that works at high pressure. When one laboratory finally succeeded, it was very much on the basis of trial-and-error, with many of the parameters of the successful device not fully understood. The essence of the argument then is that the transfer of practical knowledge, of a sort not typically contained in journal articles, was crucial to the process by which other groups subsequently learned how to do it. In other words this tacit knowledge played a crucial role in successful replication. Unlike Collins's experimental physicists, the scientists in this book were rarely looking precisely to replicate what another had done. For example, Ledley's response to the EMI scanner developed by Hounsfield, or the responses of Mallard and of Andrew to Lauterbur's announcement of the principle of zeugmatography, do not correspond to what Collins describes. Andrew and Hinshaw thought "there must be better ways of doing it," while Ledley set out to develop a device both cheaper and with broader medical application. Unlike experimental physics, sociocognitive structures here seem to permit (or encourage) less focused replications, broader sets of problematizations in which both formal and tacit knowledge may well be sought elsewhere. Moreover highly salient professional and industrial interests give these structures something of the character of the Starnbergers' "hybrid community." Patent protection makes it essential that any replication that is to compete (professionally and commercially) differ in essential respects. Speed in developing a comparable device will often be crucial, and the cost at which it can be produced will always be an important consideration. Patterns in the activities, in the various sequences of problematizations through which the technologies are constituted, are associated with structures corresponding more to the notion of a hybrid community than to the usual idea of a scientific discipline.

Careers in Diagnostic Imaging

In this section I discuss three model careers exemplified in the case studies. The first, the *conformist career*, is dominated by the shared interests that bind the interorganizational structure together. The second, the *nonconformist career*, is made wholly outside this structure, challenging established interests in x-ray technology. In discussing these two, the task will be three-fold: to establish the structural characteristics of the evolving career; to indicate the sequence of problematizations corresponding to these structures; and to try to make clear the process of mutual shaping between research and changing structure. It is there, in essence, that the work of constitution is done. The third sort of career, and the most complex, I call the *contested career*. It is dominated by protracted attempts on the part of established interests to gain control of a technology that, because it arises outside the sphere of those interests, is potentially threatening. Here we shall have to characterize these attempts and the strategies adopted and try to clarify their significance for the work of innovation-oriented research.

Conformist Careers
This notion refers to the constitution of a new imaging technology wholly under the influence of, and in relation to, the interorganizational structure centered on the manufacture and use of x-ray technology. How and why should innovation of this sort take place? Economists' discussion of the relations between monopoly power and innovation is an appropriate starting point. Schumpeter is the best known advocate of a positive association between monopoly power and innovation (Schumpeter 1975). Among his reasons for believing that monopoly stimulates innovation is that the monopolist can be far more certain than any other firm of being able to reap the benefits of innovation. The likelihood of commen-

surate return is thus the greater. Others have argued that firms in possession of a monopoly may see little reason to innovate, preferring to try to hold on to the monopoly they already enjoy. Admitting that both kinds of influences operate, Kamien and Schwartz have argued convincingly that a market structure intermediate between monopoly and perfect competition should generate the highest rate of innovation (Kamien and Schwartz 1982, 104). Since x-ray producers have a partial monopoly, and provided one ignores the structure of the user sector, this reasoning ought to apply. Do we therefore expect the x-ray industry to be innovative? We know that it always has been, and we have looked at some of the reasons. The present analysis is more complex, since it refers to the structure of the user industry as well as that of the producer industry, and it seeks to explain not just the rate of innovation but the *nature* of the innovations that arise as well. The history of the field is marked by a continuous process of marginal technological improvement. This manner of progress seemed to respond best to both industrial and professional interests.

Within the framework of this book, the point about a career of this sort is an extension of one made by economists in support of the Schumpeterian hypothesis, namely that monopolists are best able to fund the work of innovation (Kamien and Schwartz 1982, 28). The point now refers to the capacity of the existing interorganizational structure to provide not only the economic but also the social and cognitive resources needed. If we look within this structure, at the more tightly knit collaborations where the work of constitution is being pursued, we find something like the following four phases.

Exploratory   In the exploratory phase work is first concentrated in a single collaboration between a radiologist and an established manufacturer of roentgenological equipment. This might arise in various ways: by a radiologist starting work, possibly with a physicist or engineer, and then approaching a firm with which contact exists; or work beginning in the research laboratory of a firm, which then rapidly draws on its existing contacts to find a research collaborator in academic radiology. Increasingly, given the speed with which knowledge of the project is likely to diffuse (Mansfield 1985), it is likely that other collaborations involving competitor firms and other radiologists will very rapidly be set up. There is likely to be little resistance to the project—to the contrary—and since social, financial, and cognitive resources are present in abundance, progress is rapid.

Development   In the development phase the central locus shifts to industry, with various competing firms each working with a number of radiological departments. Firms from outside the industry are unlikely to attempt to enter the competition; were they to try, entry barriers of various sorts could be erected. Within a relatively short time a number of prototypes are in existence. At this point clinical data demonstrating the utility of the device in general and (more particularly) of each individual model become important. Radiologist-collaborators begin to mobilize support for the device (immediately forthcoming) and for the product. This involves actively publishing in the radiological literature.

Assessment   In the assessment phase the locus of problematization is diffused throughout the organizational structure. Sales are growing rapidly, while academic radiologists are engaged in establishing the precise use they wish to make of the new device. Initial growth of the market, however, is largely independent of the outcome of their evaluations. Rather it is determined in the first instance by whether or not leading research hospitals view the device as potentially status enhancing. If they do, other potential purchasers will be inclined to follow their lead. Competition between manufacturers will take place according to well-known principles or agreements. The behavior of hospitals, and so the market, is likely to be regulated by government legislation or funding constraints. This could slow growth or lower the ultimate plateau. Substantial research publications, detailing uses of the technique, will appear in the radiological literature. The interorganizational structure clearly functions as a reputational community, in the way that van den Daele et al. denoted by the term "hybrid community."

Feedback   Feedback of information regarding desirable modifications takes place with considerable facility and speed. Because lines of communication between radiologists and manufacturers are well established, and embody mutually acknowledged principles of reciprocity and common interest, the feedback process is unhindered. It is likely to begin early—conceivably within the development phase, with development work proceeding in parallel with setting up manufacturing capacity. This situation corresponds to what von Hippel calls a "customer active paradigm."

The case study materials yield nothing precisely corresponding to this picture. The nearest we have to it is the CT study, which differs from the ideal model only with regard to the exploratory phase. After suggestions in the literature from Oldendorff (a neurologist) and Cormack (a physicist) failed to be picked up, it was in the research laboratory not of an x-ray manufacturer but of a firm with almost no experience of medical markets that work was taken further. The crucial collaboration between an industry-based electronics engineer and a radiologist did not emerge from established relationships but was brought into being through the agency of a government department. Very soon thereafter—within four years of the end of the exploratory phase—the career structure had assumed the form detailed in the model.

What is the sequence of problematizations corresponding to this? Recall that EMI, perhaps thinking in terms of large markets, or back to its earlier experience with thermography, suggested a device for mass screening. The DHSS and their expert advisors had other ideas. Their suggestion that a device for neuroradiological purposes would be preferable structured Hounsfield's work during the late 1960s. A final decision was made early in 1970. Despite EMI's outsider status, considerations that played a role were clearly influenced by the norms of the x-ray community, including radiologists' (long-acknowledged) need for a less invasive method of brain imaging. Moreover, a device for mass screening had to meet more stringent cost norms than did a device for neuroradiology. From this decision to go for brain imaging followed the need to establish a working collaboration with a neuroradiologist. From 1970 until the end of the exploratory period in 1973, work was dominated by the specific implications of and interests in brain imaging.

Radiologists are expert in the interpretation of visual representations, and they attach a good deal of importance to the quality of those representations. For Howry, looking to develop ultrasound as a technique for general radiological investigation of soft tissue, producing pictures of sufficient quality had been a major objective. For Hounsfield, the computer reconstruction technique meant that diagnostic data could be presented either in the form of (reconstructed) pictures or of digitalized data. That visual images would have to be retained as an option, whatever the additional information content of quantitative data, was soon clear. It was the visual representation, not its possibly data-richer digital basis, which was crucial.

During the subsequent development phase, work in the many locations could take place on the basis of rapid and widespread consensus. It was clear what potential uses of the brain scanner within neuroradiological practice might be; Ambrose had sketched them out in his 1973 paper. It was equally clear on what criteria the details of its utility would be established. From this consensus emerged the vital characteristics of the development phase: the certainty that radiologists would regard increased speed of imaging, and improved image quality, as most desirable improvements to the existing EMI scanner. The possibilities of extending applicability from the head to the whole body also promised professional and economic reward. The indications for further work were clear. Together with the certainty of the market, this consensus was the source both of the ferocious competition and of the rapidity with which improved model followed improved model.

Just as competing development programs followed almost instantly on EMI's initial announcement, so assessment could also begin enthusiastically, rapidly, and consensually. Diffusion of the brain scanner (and subsequently the whole body scanner) was rapid, limited mainly by manufacturing output. In the neuroradiological area progress through the clinical research agenda was also rapid. How confidently and how accurately could specific brain pathologies be diagnosed, relative to other techniques then in use? How could CAT scanning be deployed so as to reduce the need for invasive techniques like pneumoencephalography? As the new scanner rapidly took its place in the clinic, so that patients to be scanned were no longer selected according to research protocols, new and different problematizations soon followed. Reflecting on accumulated experience, radiologists started to consider the implications of the CT scanner for the overall social and technical organization of their practice. Questions that emerged in the professional literature included the following. What effect does availability of the technique have on use of other procedures? What does the scanner cost to run, and how intensively can it be operated?[2]

Structural characteristics (size of the market, profitability of the technology, the determination of the x-ray multinationals to recapture markets and regain reputations) meant that the feedback phase started quickly and rapidly achieved breakneck speed. This continued until U.S. government controls on market growth started to bite. Trading on their names and long-established relationships, as well as their enormous R&D resources, the radiology companies could force the hands of their less-experienced competitors. Upping the ante in the development process was one way to do this. There was little doubt as to what the objectives were to be; the feedback process was contiguous with the development process. The progress from first-generation scanners taking five minutes to complete one scan to third-generation scanners taking five seconds took only five years to achieve.

Successions of problematizations are structured by the social and cognitive configurations in which they emerge. The notion that in the assessment phase of CT scanning, for example, research derives from the concerns, cognitive and social resources, and conventions of the x-ray community is of course central to the argument of this book. But there is another side to the coin. The questions posed in research, the solutions proposed, must at the same time contribute to *advancement* of the technology. How is that duality achieved? How *was* it achieved in the CT career?

Consider the work James Ambrose reported in his first CT paper of December 1973. Ambrose had to explain what was unusual about the prototype device: how it was operated, how the patient was to be positioned, the consequences of the fact that it generated numerical data. The exposition was firmly rooted in the norms and assumptions of professional practice. There was a display of sections made through a normal brain removed at autopsy, with textual guidelines for viewing the displays. Exploration of the technique was rooted in neurological practice, which defined the "common lesions which require to be identified." After discussion of the technique, and of the interpretation of its representations, a "small series of seven patients with chronic subdural hematoma" was examined. On the one hand this enabled the rules for interpretation to be further articulated. But on the other hand, since these were real patients, the reader was in effect being asked to consider the technique in relation to his or her own clinical practice. In the Discussion section of the paper it is this practice, and the potential place of the scanner within it, that provided the focus. Ambrose's speculations regarding how the device might come to be used ("like radioisotope brain scanning, it will undoubtedly come to occupy an increasingly important position, and in similar fashion will no doubt, supersede contrast radiological procedures in the investigation of many patients"), and how it might be improved technically, deliberately pointed toward the next phases in the career of CT. That so much could be accommodated in a single paper is remarkable. The point is that Ambrose could rely on, and need not build, a consensus over the desirability of the technology. Within a few months larger clinical series and statistical analysis followed from other centers in possession of the first prototypes. The assessment of CT scanning was underway.

Nonconformist Careers

A starting point here might be the argument advanced against Schumpeter, to the effect that firms lacking monopoly power have the greater incentive to innovate, with far more to win and far less to lose. Dorfman's work suggests that such an argument is too simple to explain actual patterns of innovation. She points out that while IBM held on to its mainframe market, when a market for microcomputers emerged other companies took charge of it (Dorfman 1987). Firms that dominated the vacuum tube market failed to take control of the transistor partly because they failed to see it as anything other than a smaller substitute for their existing product. Both examples entail reference to characteristics of the specific technologies. In this sense they are an advance over theories of the *general* advantage (or disadvantage) of monopoly power. However the explanation here cannot wholly parallel Dorfman's account of the transistor, in which failure of industrial vision plays such a vital part. If we need to invoke failure of vision (as some participants do in the case of thermography), then it has to be a failure of collective vision, for which suppliers alone cannot bear the whole responsibility.

We can imagine a nonconformist career of this sort initiated by a solid-state physicist or a materials scientist (for example) looking into the medical relevance of what he knows. He might be motivated by sheer curiosity, or by illness close at hand in his family. We can also imagine a medical scientist seeking help with a problem on which she is working. Is it possible to perform such and such a procedure more effectively? Is there a new material with particular sorts of properties? Early collaboration involves a medical specialist and a technical specialist. At this stage, or a slightly later one, a firm with expertise in the technology in question is approached. If barriers to entry are low, if the firm can redeploy its existing expertise without considerable expense or uncertainty, it might be tempted to develop a product for the medical market. Crucially, these collaborations and negotiations are not sustained by existing interorganizational relationships. For all those involved, they present major uncertainties and challenges as much social as technical. The history of a technology of this sort readily lends itself to the historian or sociologist inclined to exaggerated emphasis on the "heroic molding of nature and society."

If we set out the structural characteristics a career of this sort might show, they are likely to be phases like the following.

Exploratory   In the exploratory period the basic element is a collaboration between a medical scientist (probably, although not necessarily, a nonradiologist) and a technical expert, who may be based in the university or in industry. This may have arisen totally by chance, although mediation by some sort of transfer bureau or government agency is quite conceivable these days. In any event, the collaboration has something of the unfamiliar about it for both partners.[3] If news of what is going on gets around (whether via medical or technical circuits), other collaborations may come into being in other centers. By definition these involve equivalent technical expertise, but medical collaborators could conceivably be from other specialties (perhaps including radiology).

Development   In the development period the locus of work shifts to industry. If barriers to entry are low enough firms with requisite technical expertise, having been attracted by the possibilities of new markets, begin development work in conjunction with their medical collaborators. Uncertainties are much greater than in the conformist case, and subsidy of R&D costs might be important. These days, given the many schemes aimed at transferring knowledge from the university to industry, the establishment of one or more spin-off firms is also a possibility. If medical collaborators are drawn from more than one medical specialty, then the various prototypes being developed might differ significantly from one another in terms of the uses for which they are designed. Numerous publications in the medical and scientific press serve to arouse interest in advance of production. A market or markets must be established, and that depends upon convincing medical opinion that the new device is interesting and potentially useful.

Assessment   The assessment phase depends on the emergence, or not, of alternative parallel forms and segmented markets. Firms might at this stage be tempted to try to extend uses into various specialties, hoping to stimulate as yet unanticipated uses and markets. Markets grow slowly to begin with, awaiting preliminary assessments in the literature, but we cannot easily generalize about their subsequent growth. In contrast to the former case, it cannot be taken for granted that specialists will see the possibility of building scientific reputation on the basis of early publications relating to use of the technique. If distinct markets are emerging, corresponding to distinct specialist designs, uneven growth among them leads firms to consider shifting from one to the other. The constitutive work here is located principally within the user community. Each specialty, if there is more than one, assesses the relevance of the new device for its own particular practice and according to its own particular standards, conventions, and needs. Publications, which to begin with appear principally in the journals of the specialties concerned, might ultimately cohere in one or two dedicated to the technology. A successful technology might provide the basis for a new reputational structure.

Feedback   Feedback processes do not take place through long-established and familiar channels, as in the conformist case. The notion of customers as the major (tried and trusted) source of innovations, which lies at the heart of von Hippel's notion of the "customer active paradigm," is less applicable here. Combined with the fact that applications for the technology have to be established before substantial interest in it emerges, this means that feedback processes, aimed at improving available models of the technology, will not be initiated so rapidly.

Of the empirical case studies, diagnostic ultrasound most nearly approximates this career pattern, although thermography and magnetic resonance imaging resemble it in their early stages. In order to understand the kinds of problematizations associated with these structures, we therefore refer principally to the ultrasound material.

In the absence of any integrative structure, collaborators must initially draw cognitive and social resources from the distinct disciplinary or professional communities in which they work. Uncertainties are considerable. Complex problematizations result from attempts to construct a new discourse, to build a new research agenda, while drawing resources and legitimacy from diverse communities. The problematizations we can distinguish in the case study material show clearly the importance of *building*, rather than simply addressing, an audience; of *constructing*, not working through a research agenda; and of continuing to seek legitimation from the original disciplinary or professional community.

For the pioneer, engaged with the vision of a radically new technology, early publications play an important signalling function. In the cases of both ultrasound and MRI many papers were published long before a commercial instrument was under development. At a very early stage scientists tried to draw the attention of a wide segment of the scientific community to the new connection between medicine and technology they were exploring. Journals like *Science* and *Nature*, with their high status and wide readership, have an important function in that respect (Myers 1985).

Here are the MIT/MGH workers presenting their early transmission approach to ultrasonic investigation of the brain in *Science*: "This report deals with the initial progress of a long range program on the application of ultrasonic techniques to medical problems. An immediate goal is the detection and localization of intracranial tumors and cerebral anatomic abnormalities" (Ballantine et al. 1950). Wild and Reid, presenting their reflection method two years later, set about connecting two discourses, two communities, in a similar way:

The results of preliminary studies on the use of a narrow beam of 15 megacycle pulsed ultrasonic energy for the examination of histological structure of tissues have been sufficiently encouraging to warrant the development of the apparatus that is the subject of this report.

The record shown at H is believed to be the first two-dimensional echogram of biological tissue to be recorded. (Wild and Reid 1952a)

Similarly, Damadian wrote in *Science* that "NMR methods may be used to discriminate between two malignant tumors and a representative series of normal tissue," suggesting that NMR combines "many of the desirable features of an external probe for internal cancer" (Damadian 1971).

Scientists seek legitimacy and credit for what they are doing in their research community of origin. This they do by problematizing aspects of the technology in terms of the theories, methods, and concerns of that community and by publishing the results in its journals. Thus Wild, a surgeon, very rapidly applied his ultrasound technique to characterization of the spatial characteristics of tumors. However these disciplinary significances are more clear-cut in the work of physicists: for example, the MIT Acoustics Lab people in the early 1950s; or Andrew, Hinshaw, and Mansfield twenty years later. For the Acoustics Lab scientists Hueter and Bolt, it was important to root design of their device for transmission imaging of the brain in as comprehensive as possible experimental studies. Writing in the *Journal of the Acoustical Society of America* in 1951, they discussed the design parameters on which they had worked. These included the theoretical problem of the relations between resolution and ultrasonic frequency, size of transducers, and relations between attenuation of the beam by different substances and frequency used. The questions for magnetism experts two decades later were of course quite different, but the underlying approach shows pronounced similarities. Hinshaw, writing in *Physics Letters* (1974) and in the *Journal of Applied Physics* (1976), provided not only description but also theoretical analysis of the method of image formation he had chosen. He too discussed its advantages over other competing methods in terms of experimental parameters; for example, the uniformity of the magnetic field or the computing resources needed.

By the end of the exploratory phase there must be good reasons for belief both in the technical feasibility and in the medical applicability of the technique. What precisely this entails depends upon whether or not the starting point is rooted in medical practice. However in all cases interaction between the technical choices and trade-offs involved in design and the assumptions made regarding possible medical applications becomes increasingly significant. As experience accumulates, the medical and the technical areas are gradually brought into alignment (Fujimura 1987), tasks and commitments in one area having increasingly salient implications for those in the other.

For Wild in the early 1950s, as for Damadian and Mallard two decades later, the medical objective of the work derived from reflection on clinical practice. The focus was cancer diagnosis, and in each case there was a clear attempt to get "a working machine for clinical use as quickly as possible" (Wild). This conviction guided development work and led to early concern with establishing diagnostic criteria. Wild, the practical surgeon, opted for a simple phenomenological criterion based on comparing areas under traces. Damadian, the laboratory researcher, opted initially for a criterion rooted in prior work on the biochemical correlates of pathology. Later, with that work under pressure, he modified his criterion, identifying one that simply worked, "whatever the cause." The progression of Wild's work shows the alignment between the medical and the technical. On the basis of his early experience with U.S. Navy equipment, Wild chose to work at a frequency that, while it maximized resolving power, resulted in low penetration. This restricted medical applicability to accessible sites on the body. Wild consequently switched his attention from the brain to the female breast, and it was on the basis of breast studies that he defined his diagnostic criterion. Subsequently abandoning his linear traces for a new device producing two-dimensional images, Wild emphasized the advantage of being able effectively to *locate* as well as diagnose the tumor. However the new form of representation in turn led to the need to redefine the diagnostic criterion.

For nonphysicians the process of aligning technical and medical tasks and commitments is different. To begin with Mansfield, Andrew, and Hinshaw had no medical objectives: these began to play a role only as they emerged from reading and discussing. This made for a more cautious and gradual approach (which provoked Mallard's ironic observation that "I'm not interested in fingers and apples and oranges"). The work showed a growing orientation toward specialist communities having unfamiliar concerns and priorities and different conventions (see figures 7.1, 7.2). Unfamiliar considerations, with no obvious rationale in physics, gradually became relevant. For example, scaling up the apparatus, though unnecessary while the sole object was to compare possible methods of image formation, was seen to be essential if the technique were ever to have medical application. It soon became apparent that increased sample size meant that producing an image took longer. For a physicist this was not necessarily a problem. However within a medical context overlong imaging time is likely to be problematic in one of a variety of possible ways. It was for medical, not physics, reasons that Hinshaw and Andrew began to consider faster imaging techniques.

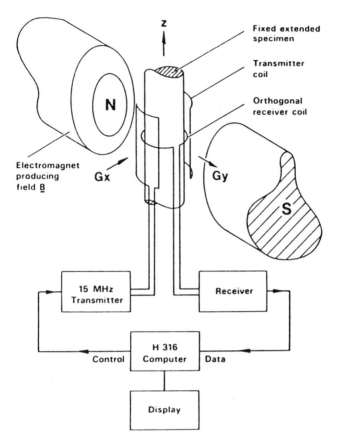

Figure 7.1
Schematic representation of an NMR imaging instrument published in the *British Journal of Radiology* in 1977. From Mansfield and Maudsley 1977, 189. Reprinted by permission of The British Institute of Radiology.

We saw that Hounsfield and Ambrose's announcement of the arrival of CT scanning provoked rapturous enthusiasm and ferocious competition. In the case of a conformist career the extent of the competition derives, in part, from shared preconceptions regarding the desirable characteristics of an (improved) scanner. Development and feedback aimed at further improvement are scarcely distinguishable. The beginnings of the development phase of a nonconformist career are very different. Neither the first ultrasonic images nor Damadian's whole body image of 1977 were greeted with any great enthusiasm. In both cases the reactions

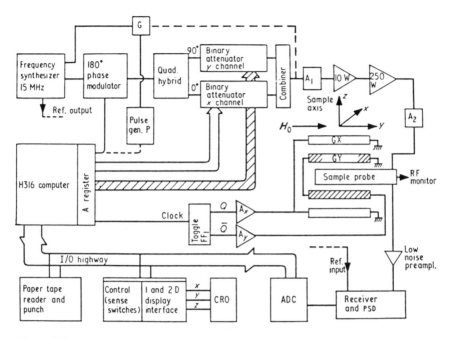

Figure 7.2
Schematic representation of an identical instrument published in the *Journal of Physics (E)* in 1976. From Mansfield, Maudsley, and Bains 1976, 273.

outside the groups concerned were slight, with competition building up gradually, and the transition to a development phase less clear. In neither case was development governed by a sense of agreement either over the range of problems to which the technique would have relevance or over the criteria on which it would be assessed. In the MRI case this uncertainty, as much as corporate strategy, shaped industrial development programs.

There are various indications of the importance of this uncertainty regarding essential features. Very large firms that had the necessary resources chose to proceed in parallel with alternative systems. EMI, influenced by their own experience with the CT scanner six years before, decided once more to aim at brain scanning. The sequence of problematizations was nevertheless significantly different from that which shaped CT scanning. The first brain image was made public at a very early stage, signalling EMI success. In the two years between this announcement and the installation of the first prototype in a hospital, major changes were made to the design. The device installed at the Hammersmith Hospital in January 1981 had a different magnet system and, most importantly, was a whole body scanner. Whereas Ambrose's first paper had set out the lines of future assessment and development at the outset, early EMI/Hammersmith work was much more speculative as to possible imaging technique and uses.

Marked by continuing uncertainties, the beginnings of MRI development are clearly different from those of CT scanning. Ultrasound is different again. In the MRI case each group confronted the same uncertainty over what the principal uses of the technique might prove to be, and each developed its own strategy, based on expectations and interests. Twenty years earlier ultrasound groups did not confront the same global uncertainty. A simple version of the technology was readily available. The groups active in the development phase of diagnostic ultrasound looked at the technology in terms of their distinctive specialist concerns. In some cases this yielded local certainty. Lecksell, the Swedish neurosurgeon, knew exactly what he wanted; Edler, his cardiologist colleague, began with various possibilities in mind. Ultrasound was taking distinct specialist forms, while MRI was not. Thus competition between the two Nottingham physics groups (Andrew's and Mansfield's) was feasible in a way that was inconceivable for the Lund departments involved with ultrasound, where distinctive medical commitments were of paramount importance. The fact of these distinct forms, and the (partially) segmented market that followed, had major consequences for the assessment of ultrasound. For example, the fact that the technique was effectively displaced from neurological practice by the advent of CT scanning had no effect on its growing use in obstetrics. The partially segmented character of diagnostic ultrasonics is reflected in a research program like Kossoff's, which sought to contribute both at the level of the technique as such, and at the level of the individual specialty. It is reflected also in the growth of a reputational community largely separate from that of radiology. Medical ultrasound has its own international associations and seven international journals (listed in the *Science Citation Index*), many of them cited more frequently than the major radiology journals.

In a nonconformist career, too, the succession of problematizations has a dual character. Research both reflects changing social and cognitive configurations and at the same time contributes to advancement of the technology. In the CT case we probed this duality by focusing on published work associated with the transition from exploratory to development and assessment phases. In regard to the nonconformist career, as demonstrated in the case study material, two intriguing questions arise. How does lack of consensus, characterizing both ultrasound and magnetic resonance imaging, manifest itself? Second, can we see how it was that ultrasound (unlike MRI) took the segmented form that it did? How did research both reflect and contribute to this branching career path?

The first major report on EMI's magnetic resonance imaging project (Young et al. 1982), with its eleven authors, was very different from Hounsfield and Ambrose's back-to-back accounts of the same company's CT scanner. Despite its later place in the career of the technology, the Young paper displayed a number of ambiguous features. On the one hand, as we should expect, it was more firmly located in the clinic. For example, it reported on fifty patients (compared with Ambrose's seven). The reference for interpreting images was no longer a photographic section obtained at autopsy but clinical images obtained with the now familiar technique of CT scanning. Some indications of possible advantages over CT scanning were offered. In other respects the paper pointed back to the exploratory phase of the technique. For example, it reported not only on a clinical series of fifty patients but on thirty volunteers as well. There were detailed accounts of the principles of the technology and of the different imaging sequences available. The images provided by these three sequences not only reflected distinct biochemical properties, but they also looked different, so that a key to each had to be provided. The body of the paper consisted of examples of normal appearance and a few clinical cases, some of which had a distinctly experimental character. For example, imaging of the heart was attempted only with ten volunteers and included an experiment on the appearance after inhalation of 100 percent oxygen. Where the discussion section of Ambrose's paper had pointed firmly forward, the MRI paper showed a pronounced ambiguity, hinting at clinical possibilities while stressing the need for more work:

The spin-lattice relaxation time T_1 is a sensitive but relatively nonspecific diagnostic parameter that may prove useful for quantitative assessment in diagnosis and monitoring therapy.

The variety of sequences possible with NMR allows a choice of image to reflect varying degrees of proton density and blood flow, as well as tissue spin-lattice and spin-spin relaxation times. While this gives NMR considerable versatility, much more work will be required to establish the optimum technique for particular conditions and circumstances. (Young et. al. 1982)

Embodying as they do the branching career of the technology, very few of the original papers published on ultrasound look across the regions of the body in this way. How does branching manifest itself? Donald, MacVicar, and Brown's 1958 paper in the *Lancet* marks the origins of gynecological and obstetrical ultrasound. Recall that Donald had come into contact with Kelvin Hughes, makers of an ultrasonic flaw detector, at the time the firm was encountering some interest in its product among neurologists. Donald had borrowed one of these devices and tried it out on a few of his patients. Together with a brief account of the general principle of ultrasonic echography, that was the starting point of the *Lancet* paper. Like the Hounsfield/Ambrose papers, this paper, too, covered a lot of ground. Studies of the possible harmful effects of ultrasound, carried out by a colleague from the anatomy department, were described at length. Involving examination of brain tissue from newborn kittens, these were the sorts of experiments we are inclined to associate with the exploratory phase of an imaging technology. It soon became apparent that the flaw detector worked only on "the simplest reflecting interfaces" and was inadequate for investigation of the abdomen. Discussion of the more sophisticated instrument developed by Tom Brown led the reader toward the development phase and at the same time into the clinic, with the new apparatus mounted on a standard hospital bed table. The examination technique was described in terms specific to abdominal examination. It involved smearing the patient's abdomen with olive oil to exclude air and ensure acoustic coupling. The probe was then "moved slowly from one flank, across the abdomen, to the other flank. . . . Several cross-sections are taken thus at different levels between the symphysis pubis and the xiphisternum." With the results section of the paper the reader was firmly in the clinic. Results had been obtained from 100 patients, "mainly . . . the routine clinical material of our own department." Like the MRI paper Donald ended on the note of cautious optimism that seems to be characteristic of the nonconformist career. Like Ambrose, Donald devoted much of his paper to representations of the various conditions with which the specialist was confronted (though Donald did not include sections or the results of other imaging procedures for comparison). Despite this formal similarity, Donald, MacVicar, and Brown's paper differs from the CT papers in one vital respect. The Glasgow group carefully distinguished their technique from those of other ultrasound researchers (Wild, Howry, the cardiologist Effert) in terms of differences in clinical interests. It was by *assuming* medical and technical specialization from the start, by studying the technique in relation to the clinical material of their own specialist practice, and by restricting their speculations regarding its possible utility, that Donald, MacVicar, and Brown contributed to the branching career of ultrasound. Subsequent Glasgow publications increasingly assumed the utility of the new technique, focusing on its precise relevance for the distinctive concerns of obstetrical practice.

Contested Careers

In opposition both to the Schumpeterian and anti-Schumpeterian positions, some economists have argued that a firm having monopoly power is best placed to be a "fast second" in the development of an innovation (Baldwin and Childs 1969). A number of empirical studies have concluded that large industrial laboratories tend to produce minor inventions, with major inventions emerging from less dominant firms. Kamien and Schwartz summarize the point: "Because of its resources and established reputation and channels of distribution, a firm with monopoly power can afford to wait until someone else innovates and imitate it quickly if it appears to be successful" (Kamien and Schwartz 1982, 30). This is more or less what happened with the CT scanner. We saw how EMI, having invented CT scanning, first lost its monopoly and was ultimately driven from the market. This can be explained in terms of the mobilization by established firms of the social resources to which Kamien and Schwartz refer.

Though this "fast second" hypothesis is extremely plausible, it is also overly simple. In the field of diagnostic imaging existing interests are not defended only by (oligopolistic) producers, and analysis cannot be restricted to economic markets and entry barriers. Radiologists defend their prerogatives, when challenged, no less than industry defends its markets. When the *Lancet* dared to wonder whether, given their "specialized viewpoint," neuroradiologists and their neurological colleagues were "necessarily the best people to advise on the strategic deployment of the EMI scanner," there was a scandalized reaction. When Creditor and Garrett, writing in the *New England Journal of Medicine* in 1977, wondered at the adequacy of the information base on which $700 million had been committed, they too received a literary beating. Moreover defensive strategies mounted by industry and the radiological profession are not independent of one another.

The second point derives from a difference of emphasis from both Kamien and Schwartz and the authors they cite. The argument of this book is not that monopoly producers can afford to wait (though the CT example suggests that they can), but that under certain circumstances they feel themselves compelled to act. Organizational theorists, discussing responses to growing uncertainties in an organization's environment ("turbulence"), often make a similar point, and the notion of "domain defence" (Miles and Snow 1978) has been introduced. Carney (1987) has discussed the types of collective strategy firms in an industry may adopt when faced with the "invasion of new organizations or technology." The perception of threat provokes defensive responses from organizations, and these typically involve collective action.[4] The argument here is that defense of the collective interest, of the "field" or "domain," involves seeking to assimilate the threatening technology to integrated structures both of production and use (or, conceivably eliminating it totally).

There is no model career here since everything depends on when, and how, conflict develops. What strategies are available in the exploratory, development, and assessment phases of career? What kinds of problematizations do they entail?

Exploratory The basic element in the exploratory phase, as we have seen, is a collaborative relationship between a medical researcher and a technical researcher (physicist or engineer). At some stage, through one of a variety of processes, industry becomes involved. In the conformist case this central collaboration was located within the interorganizational structure based on x-ray; in the nonconformist case it was not. Conflict will emerge at this initial stage only when collaborations of both kinds are present. Yet the existence of competing visions of the future of the technology is not a sufficient condition for conflict. It is an (empirical) question as to whether the distinct social and cognitive resources on which the two (or more) collaborations draw affect the interaction between them. Conscious awareness, and defense, of structural interests (economic, professional) is yet another matter.

Development In the development phase the locus shifts to industry. Conflict emerges here if, through capitalizing on special technical advantage, firms from outside the industry manage rapidly to establish themselves, possibly in partnership with nonradiologists. To establish its position, such a firm will have tried to erect entry barriers: making use of patent protection, trying further to advance its technical expertise, and attempting to build up a customer base. Lacking existing ties into the medical world, these firms typically investigate various possible medical markets, perhaps by making instruments widely available on loan. Rapid publication of favorable reports in the literature is vital. If the device looks promising, x-ray manufacturers, working together with their radiologist clients, attempt to drive the competitors from the field. To overcome the entry barriers, they must first invest sufficiently in R&D to attain technical parity. Subsequently, maximum use will be made of existing assets: reputation, sales and service networks, and (conceivably) the ability to engineer unique compatibility with existing products. Feedback processes may also play a role, as R&D stakes are gradually raised as a means of driving out, or inhibiting, competition. Kamien and Schwartz show further that a producer's known ability to improve its product in the event of imitation discourages would-be imitators.

Assessment Users, not producers, play the leading roles if conflict is still taking place, or emerges, at the assessment phase. The struggle for control will take place both at the level of the individual hospital and in debates within the medical profession at large. These struggles may take various complex forms, for broader issues may be invoked. For example, if the technique is sufficiently expensive, segments of the medical community affected only indirectly, through the financial ramifications of the device, may become involved. Arguments about the further technologization of medicine might involve still more distant interest groups. Government regulatory requirements have added still further to this complexity. By imposing a framework within which the safety, costs, and utility of the technique are to be established, government regulation may change the rules of the competition in favor of one or other group of participants.

Successive career phases, with their characteristic configurations of actors and their particular loci of constitutive activity, suggest strategies of challenge and conflict characteristic of laboratory life, economic life, and professional life respectively.

What are the strategies open to competing laboratory-based coalitions? These are likely to involve imposing new consensus regarding what counts as a significant contribution, as an effective technique, or as a working design—"increasing the material and intellectual requirements" (Latour and Woolgar 1979, 119–124). Latour and Woolgar showed that, by securing agreement to a new and more rigorous set of criteria of proof, entailing new investments in instrumentation, a research group was able to drive competitors from the field. In the language of this book, we would then see a shift in problematizations attendant on the deployment of these new resources.

In the cases both of ultrasound and thermography we see research approaches based in radiology (Howry, Gershon Cohen) and other projects led by cancer surgeons (Wild, Lawson, Lloyd Williams). Are relations marked by attempts to drive some among them from their respective fields, in the way Latour and Woolgar describe? If so, are these attempts to be interpreted as the defense of radiological interests against potential threat?

In both cases there are hints of systematically different perspectives associated with radiological and nonradiological approaches. Howry rooted his claims on behalf of his first ultrasonic images in the adequacy with which they represented anatomical structures, whereas Wild rooted his in statistics based on applying a simple diagnostic criterion. In stressing that thermographic assessment is fundamentally different from "anatomical alterations that represent the cornerstone of breast cancer diagnosis," Lawson was distancing his work from radiology. There was a clear difference in tone between the thrusting and entrepreneurial style of Gershon Cohen and Barnes on the one hand and the cautious optimism (characteristic of nonconformism) that marked Lloyd Williams's presentations on the other. To attribute these differences to intentional conflict based on opposed interests is to claim too much. It can plausibly be argued that Howry and Wild actually saw themselves as doing different things. Furthermore there are clear similarities between the diagnostic criteria used by the cancer surgeon, Wild, and by the radiologist, Gershon Cohen. Certainly we cannot rule out the operation of competing interests. Wild did leave the field, and history has been suspiciously ready to write off his work as based on a mistake. But the case is not a strong one. The contrary hypothesis—that their relations were just as much the embodiment of common enthusiasm for a new technology, common interest in its success—seems at least as plausible. As commitments build up, as commercial interests become more significant, this ceases to be true.

As the locus of work moves into industry, challenge and conflict not only becomes more salient, it also takes other forms. The economics of innovation literature alerts us to the strategies that may be important here. If rivals have entered the field first, the x-ray industry and its collaborators seek to drive them out. Price is not at issue. Firms in this industry do not compete on the basis of price, and strategy will not include offering equipment for sale more cheaply. It might include industrial takeover, or holding on to customers by building in compatibilities with other equipment, or mobilizing feedback processes through which improved models can be developed, or increasing the rate of innovation to a level at which the competition cannot keep up.

In the development phase of each diagnostic technology processes of this kind are to be found, and takeover has always been a permanent feature of the (medical equipment) industry (Hutton 1985). There is, however, a clear historical evolution. In the case of ultrasound, industrial restructuring taking place in the late fifties and early sixties (in part connected with the emergence of microelectronics) had major consequences for the development of diagnostic ultrasound. But the technology was far from being central to this restructuring, and its retarded development was more an unplanned consequence of divestiture than of acquisition policy. Although in that case, and in that of thermography, we do see specific takeovers relevant to the theme of this book (e.g., Philips took over Bofors' thermography activity), this all takes place too late to influence the development of the technologies. It is only in later years that diagnostic technologies become central to industrial strategies. The costs of development by the 1970s are enormous, and as costs build up, success becomes vital. Although it was possible for large firms (EMI, Texas Instruments) to develop and produce a few prototype thermographs without too much worry, this was no longer the case with CT scanning. Recall the Pfizer executive's observation that, with CT scanning, one either became a "full-fledged radiology company or one got out." CT scanning became a very unsuitable technology for a company looking to diversify into the medical equipment market.

Though technical advantage, deriving from prior experience with a technology, may provide the basis for entry, it is rarely long-lived. Nor are patents much protection in the medical electronics area. When John Powell, EMI's senior recruit from Texas Instruments, argued that the firm's lead in CT scanning would probably last two years, he hit the nail pretty squarely on the head. Within three years GE and Siemens, among others, had CT scanners on the market. Given their resources, the giant x-ray companies can very rapidly mount a crash R&D program, if necessary, built around a scientist brought in from a university and with the knowledge that everything necessary is at his or her disposal. It took Philips only two years to build a prototype MRI scanner, starting from scratch. Of course this technical catching up is only the beginning. It took the x-ray companies a little longer to drive the competition from the CT market, and that they were able to do so depended in no small measure on their privileged relations with the radiology profession. Efforts were—and are—symbiotic. Radiologists preferred to buy from known and trusted suppliers. By 1976 experienced observers could already see that in the North American market GE ("whose strong name will give them credibility") would mount a major challenge. Damadian, setting out to sell his FONAR scanner, convinced of its superiority, could not understand hospitals' seeming preference for MRI scanners produced by GE, Siemens, and Technicare. There is little mystery and no need to appeal to nefarious motives. We have seen that hospitals, like other purchasers of technically sophisticated products, prefer to remain loyal to known and trusted suppliers.

For radiologists to turn the tide in favor of their established suppliers requires, of course, that they themselves dominate the market. In the case of CT scanning this was true from the start. In the cases of thermography and of MRI it became true during the technology's development phase. Suppliers of thermographic equipment found a burgeoning breast-screening program within radiological practice and an interest in adding to the armamentarium. Potential MRI manufacturers found a community whose willingness to adopt the new technology did not depend upon prior proof of efficacy or utility. Radiologists provide an attractive market precisely because of their fundamental commitment to new technology. Symbiosis can work to the still further advantage of those involved. The x-ray companies drove competition from the CT market not only because of their ability to sell to radiologists. A second line of attack, in which radiologists also played a crucial role, involved upping the R&D ante. By virtue not only of their resources but also of the effective feedback they enjoy—the effective customer active paradigm based in acknowledged mutual interest—x-ray companies are better able to introduce a continuous flow of improved models to the market.

It seems as though the development phase is generally decisive, for we have no case of radiologists obliged to carry the struggle for control forward into the assessment phase. This does not mean that radiologists may not be called upon to defend their interests. Challenges to newly imposed radiological hegemony do occur, and they may evoke echoes of earlier battles. The struggle is not, of course, the struggle for markets, but for professional—and in effect cognitive—control. The assessment phases both of MRI and of thermography show radiologists fighting to hold their gains—manifest in conflict over agendas, over research protocols, and over the problematizations through which the technique is to be assessed. But because what was at stake in the two cases was very different, conflict took quite different forms.

When Damadian first argued that nuclear magnetic resonance would be useful in diagnosing cancer, he used an argument formally similar to one Lawson had used earlier. NMR, Damadian argued in 1973, offers a way of observing and classifying tumors very different from pathologists' reliance on their appearance in a light microscope. In 1981, as the assessment phase of MRI opened, Mallard argued that interpreting images on the basis of comparison with "other images derived from completely different tissue properties, e.g. X-ray absorption," would only add "confusion." Whereas Damadian's argument in 1973 had been directed to pathologists, Mallard's remarks come from an address to the Royal College of Radiology. Mallard was convinced that indication of the areas in which MRI would be of use had to come from "biological" studies of "normal and pathological tissues." Nevertheless, attempts to establish uses by systematic comparison with CT scanning, organ by organ, were already underway. In the course of the development phase radiology had established itself as the most significant customer for the emerging technology, and the agenda for assessment was being set by radiological practice.

Criticisms regarding radiologists' suitability to decide on the value of CT scanning, and of the information base on which decisions to invest in the technique were taken, had been brushed aside in the mid 1970s. Comparable criticisms regarding MRI were potentially more threatening. Not only was the newer technique less firmly under control, but with budgetary restrictions much tighter, specialties were more consciously competing for resources. Thus it was that a 1988 *JAMA* article by Cooper et al. provoked a serious and sustained rebuttal from the leaders of the radiological profession. Cooper et al. argued that the first four and a half years' assessment studies, on the basis of which decisions to adopt MRI were presumed to have been made, were fatally flawed from a methodological point of view. More dangerously than Creditor and Garrett, Cooper et al. were driving a wedge between the process of decision making through which new imaging technology was acquired and its rationalization in the medical literature. And while the editors of the two main U.S. radiology journals chose to formulate a measured defense of their journals' (and colleagues') scientific standards, some radiologists were disposed to hit back in less measured language.

The difficulties for thermography were actually more fundamental than those raised by Cooper et al. They relate to the nature of the images and the relations of the images to pathology. In the MRI case similar concerns had provided grounds for criticism of Damadian's preimaging biophysical work, which had claimed to identify tumors on the basis of their T_1 measurements. These concerns were subsequently set aside. MRI provides various possibilities of image formation, and radiologists were confident that one or the other would always enable them to distinguish similar adjacent tissues. The absolute characterization of pathological tissue ceased to be the basis on which diagnosis was to take place. In the thermography case breast pathology was identified in a rather different way. Far too little fundamental work on heat transfer had been done for absolute characterization of tumors in thermal terms to have been proposed. Here, radiologists followed Lawson's initial lead and identified breast pathology with departure from thermal symmetry. When research by cancer surgeons and medical physicists based at London's Royal Marsden Hospital showed the complex and asymmetric thermal patterns of normal breasts, the adequacy of this diagnostic criterion became hard to sustain.

The response to this challenge was quite unlike the passion with which early work on MRI was defended. Though at the end of the 1960s radiologists were largely willing to grant thermography a modest place in their diagnostic armamentarium, the technique had little glamor. Temperature measurement was far too familiar, the technology too inexpensive, the companies manufacturing it of too little importance for radiologists. If rheumatologists wanted to use it, all well and good. That was no threat to radiology. When use of the technique in breast diagnosis began to seem problematic, requiring new and firmer theoretical foundations at the very least, radiologists largely lost interest. They were prepared to give thermography up with little more than a shrug. The exceptions were those few individuals who had already invested a good deal of their own scientific reputations in the technology and therefore had much more to lose. In the intraradiological battle that followed, the thermographers developed a number of strategies in the attempt to save the technique.

Promising Instrumental Improvement Around 1970 Dodd and his collaborators from the University of Texas' M. D. Anderson Hospital were working with Texas Instruments. Having presented their new thermograph, designed to provide higher spatial resolution than any instrument available commercially, this group took an unusual step. Despite its resolving power, "the experimental scanner described in this paper," they wrote, "cannot be considered satisfactory for routine use." They went on to indicate the directions in which they were working to further improve it. This enabled them at the same time to suggest that evaluation of thermography-as-such on the basis of then-extant commercial apparatus would be premature, and no indication of the real potential of the technique.

Redefining the Role of the Technique Work published by the AEMC group in the early 1970s exemplifies a different strategy. As the symmetry criterion for diagnosis became problematic, previous claims for the authoritative use of thermography no longer seemed plausible, and in 1972 Isard wrote, "The thermogram at this stage of development of the discipline . . . can no more by itself differentiate a benign from a malignant condition than can the temperature recording by the oral thermometer differentiate pneumonitis from a necrotizing neoplasm."

The strongest plausible claim was for the value of thermography as a preliminary screening modality—a means for safely, and without risk, characterizing a group requiring further investigation by other techniques.

Developing Phenomenological Diagnosis   The strategy adopted by Ziskin and his colleagues from Temple University was comparable to the line Damadian also took. It was to develop a phenomenological diagnostic criterion, which, because it lacks biological referents, is immune to certain sorts of critique. The essence of the procedure was to establish as many as possible "features of interest," indicators that might be generated on the basis of the thermograms and that might form the basis of a diagnostic criterion. Taking two sets of thermograms, one known from biopsy to be made up of normal women, the other from women with a tumor in one breast, the best possible decision algorithm was derived. In place of a diagnostic criterion such as a temperature difference, supposed in some way to characterize cancer, we now have a decision algorithm based on the data set on which the computer was "trained." It is a property of the data set alone, and not of pathology, and can be improved only by changing the data set, not by (for example) physiological or pathological research.

Despite these very plausible strategies, the radiological profession has not been willing to look again at thermography. Despite the technique's acknowledged utility in other areas of medicine (for example rheumatology), and despite further technical advances deriving from military R&D, an air of failure attaches to thermography. Whatever the technical advantage available to a particular firm by virtue of its military work, firms have been unwilling to invest in medical thermography.

Looked at in terms of either market or professional control, it is apparent that neither in the case of thermography nor of MRI has the x-ray community been seriously threatened. If we look at the way in which these technologies function as sources of scientific reputation, matters are somewhat different. Recall how ultrasound gave rise to a reputational community in which physicians of many kinds, as well as physicists, acoustical engineers, and other specialists participate. The pages of the ultrasound journals bear eloquent witness to the vibrancy of the field. Thermography is barely visible in the periodical literature, and those working on it cannot rely on a community constituted around the technology as their principle audience or source of reputation. Medical physicists working on thermography, for example, are likely to publish principally in the journals of that field (e.g., *Physics in Medicine and Biology*). Magnetic resonance imaging is quite different, with the vestiges of a new reputational community clearly present despite radiological control of the technology. Nuclear magnetic resonance is relevant for diagnostic medicine not only by virtue of proton imaging, which forms the basis of current MRI, but also of spectral (high-resolution) NMR used by biochemists and biophysicists to investigate degenerative processes in the body. The gradual convergence between these approaches, as well as similarity in basic equipment, underlies the emergence of a new hybrid community. This presents participants with a difficult choice. Thus, of the main actors in chapter 6, one was by 1982 thinking of leaving the field. Andrew thought that it was time to "pick up again some of one's earlier interests in physics." Damadian and Lauterbur had begun new careers: Damadian (although he continued to publish on MRI) ran the FONAR Corporation; and Lauterbur moved to the University of Illinois to take up an appointment in the Department of Diagnostic Imaging. This technology, I argued, might lead to fundamental transformation in the existing field of diagnostic imaging and hence of the interorganizational structure that has hitherto dominated it.[5]

Medicine, Technology, and Society

Attempting to elucidate this domination, to show how certain interests shape the constitution of new medical technology, has cast doubt on existing ways of thinking about technological innovation in medicine. There is no support for those who see in medicine's changing technology nothing more than perfection of the instruments of domination. Neither is there agreement with those who blame capitalist enterprise, thought cleverly to have somehow turned the medical profession into a willing customer for any new gadget or preparation. A driving force, I have argued, derives rather from the historically achieved *integration* of professional and industrial interests. In that respect it is Brown's analysis (Brown 1979) that begins to approach the explanation offered here, though with the difference that I do not derive the common cause made between the emergent radiological profession and the leaders of the growing x-ray industry from common class interests. Out of a scientific interest in the new roentgen rays grew a society in which medical people and engineers, manufacturers and professors, met as fellow enthusiasts. Gradually interests diverged. Later common interests were reestablished between medics and manufacturers but on a new basis, different but symbiotic. That symbiosis, which has fundamentally structured the subsequent events discussed in this book, is distant cousin to the notion of mutual legitimation that figures in Brown's analysis.

The successful introduction of a new medical device cannot be accomplished without the cooperation of the medical profession. The devices industry, and in this case the x-ray industry, depends upon clinical reports attesting to the utility of the innovation in question. Individual firms depend upon reports attesting to the (superior) performance of their particular product. Thus insofar as business depends upon (successful) innovation, business is dependent upon the active cooperation of medical professionals. At the same time radiologists' status, their position in the social order of medicine, depends upon their mastery of the images, and the equipment, which they control. Innovation is vital to the maintenance of expertise, status, and income, and radiologists depend upon industry providing them with a flow of new technology. It is this symbiosis that explains the rapidity of the innovation process and at the same time justifies the view that technology in medicine serves to reinforce existing power relations. This critical sociological point finds support in this book, although its further working out would require a focus less on the constitution of new technology than on technology in use (see, for example, Barley 1986).

A major concern of the book has of course been the adequacy of existing theoretical approaches to technological innovation. Simple theoretical models find no more support than do simplistic ideologies. Consider the thesis regarding the relations between monopoly power and innovation, which has largely structured this final chapter. We have seen that none of the four examples of (radically) new technology discussed in this book actually originated with a major x-ray company. It is also true that the x-ray companies have nevertheless reaped much of the profit the innovations have yielded. This is consonant with the thesis that monopoly firms are well placed to be a fast second. Plausible though it is, such a thesis proves inadequate in the face of questions like: How do monopolists secure control of a technology? When do they fail? Above all, and thinking back to the sorts of questions the new sociology of technology has posed, how does their success or failure affect the form in which the technology ultimately becomes available?

I have argued that not the structure of production alone, but the structure of use as well, and the relations between the two, exert crucial influence on the innovation process. The original visions of Dussik and Denier, Oldendorff and Hounsfield, Damadian and Lauterbur became what they did—what they are—thanks to the preferences and the work of diverse actors. But these actors are interdependent. X-ray companies can be a fast second in part because of their established relations of mutual trust and dependence with the radiologists. They secure control through strategies involving concerted action with radiologists. The interest of radiologists, as of many other specialties, in a practice firmly based in science and technology, finds wide social support. Both the cultural significance of medicine and the status of its practitioners are intimately connected with the reassurance that it provides in the face of life-threatening events. But an important change has taken place in this respect. As I argued, it is now the hospital specialist rather than the family doctor who can best provide that reassurance, for the capacity to reassure has become bound up with licensed control of the awe-inspiring resources of science-based medicine. Science is crucial to the ideology of modern medicine, and must *demonstrably* provide the discoveries on which new technologies can be based. Diagnosis and treatment, too, must derive from objective scientific rules governing proper use of these technologies. That is the real significance of studies such as that of Comroe and Dripps, which sought to provide proof for the grounding of medical progress in scientific discovery (Comroe and Dripps 1981). The overt message might have been a plea on behalf of funding for basic biomedical research. The covert message was an ideological one. The dominant ideology also shows the significance of studies such as those of Creditor and Garrett and Cooper et al. By driving a wedge between the basis on which decisions to invest in new medical technology are actually made and their scientific rationalizations in the literature, such work threatens to erode the connection between medicine and science that is of central ideological importance. Hence the violence of the reaction that it evokes.

The argument of this book does not conflict with the notion of the dependence of medical technology on science. To the contrary.

But it does give that dependence an interpretation quite different from that implied in linear science-application models such as that of Comroe and Dripps. In the approach taken in this book, research has a central, but dual, place. Actors problematize the emergent technology through the medium of research. They do so in ways that both reflect the sociocognitive configurations in which the technology exists at a particular point in time and that at the same time serve to further the process of its constitution. In other words, by looking at the detail of research, we see how the social and cognitive resources available to participants (science) are deployed simultaneously to advance both scientific reputation and technological artefact. Translated into simple ideological terms, scientific reputation-seeking has become as central a motor of technological innovation as the components of ideology to which Brown referred.

For many of the scientists whose work we have looked at, the x-ray community functions as source both of reputation and of the resources—social, cognitive, and financial—necessary for the constitution of a new device. In the course of half a century this community has not only expanded but also changed. Marginal innovation, improvement to the basic x-ray technology, remains important. But it is no longer enough. Both radiologists and industry have glimpsed the possibilities of greater fame and fortune and are naturally tempted. In recent decades innovation in diagnostic imaging has become strategically important, and costs and potential benefits now loom large in the planning of some of the world's largest firms. We have seen some of the effects of this in the case studies. Certain of the respects in which the career of MRI differs from that of ultrasound are a consequence of broad socioeconomic change. Thirty years ago chance encounters could be crucial to the origins of a new development project. Today technological forecasting, strategic planning, the availability of government subsidy in the context of innovation support programs, and systematic access to university research give innovation a very different character.

Evolution in the structure of the x-ray community is to some degree a consequence of the changing roles and responsibilities of governments. Resources made available for health care, as well as priorities in technological innovation, and regulation both of industrial structure and medical devices—all of these have had unanticipated consequences for the interorganizational network based on x-ray technology. It follows that processes of innovation have been affected not only by specifically relevant priorities and mechanisms but by unplanned evolution in the interorganizational structure as well. Despite hopes vested by some in technology assessment, there is little insight into the overall consequences of government policies for innovation in medical technology. Discussion tends to focus, separately, on two areas. One is within the context of innovation policy, where medical technology is a priority area in many countries. The second is within the context of health policy, where new technology is often seen as a major culprit in rising budgets. The analysis of this book has two sorts of implications for policy, which can be spelled out.

First, it suggests that through sponsorship of innovation in one area of medical practice, diagnostic imaging, governments are in effect furthering certain professional and industrial interests. Though no one could have predicted the course a given imaging technology would take—what I have called its career—the chances of its being assimilated by existing industrial and professional interests are clearly substantial. Such assimilation serves not only to reinforce those interests (necessarily at the expense of others), but it also has implications for the form the technology takes.

Is it possible to conceive of influencing that form? This is the elusive notion of steering technological innovation. Some commentators believe that the innovation process is in fact changing. "Even in the absence of changes in Medicare's capital payment policy, the prospective payment system creates incentives for hospitals to acquire cost-saving devices. Manufacturers are already shifting their emphasis from breakthrough technological equipment to cost-saving devices and from devices used in inpatient settings to those used in outpatient settings" (Kessler et al. 1987).

Whether or not it is true, this notion hints at the answer implied by the present analysis. In the case of diagnostic imaging, we have seen that the constitution of new technologies has largely been shaped by the interorganizational structure. It follows that one approach to steering innovation would be through changing that structure—deliberately manipulating the values and the norms governing the interactions between the professional and industrial organizations constituting it. Is that a meaningful notion?

There is evidence that interorganizational structures of this sort can be fundamentally altered. Temin's study of the FDA and the

U.S. drug market is particularly pertinent (Temin 1980). In it, he shows how the relations between producers of pharmaceuticals and their customers were fundamentally changed by government legislation. First, in 1938 there was a shift "from a market to a hierarchical arrangement in which consumers would be directed—ordered—to purchase drugs chosen by people above them in the hierarchy" (p. 48). Doctors were thus placed between consumers and their supply of drugs. Associated with this was a growing symbiosis between doctors and drug companies. Subsequently, with 1962 legislation, the doctors themselves were subordinated to the government's medical experts, and still later (1970) their role was further circumscribed.

Patients, those to whom new diagnostic technologies are applied, have scarcely figured in the analysis of this book. The assumption, supported by a little information, has been that the sick (and the potentially sick) express most plaintively the general support for high-technology medicine, which is a hallmark of western industrial society. Only very briefly did the skeptical patient appear when I referred to criticisms from the women's movement of routine use of ultrasound in obstetrics. It was too soon to see if these kinds of concerns with safety, and in particular with safety *in relation to* routine (over)use, might have implications for the further development of the technology. They certainly led to renewed research attention for the effects of exposure to ultrasound, questions that had been set aside as the technique entered routine use. Were patient organizations systematically to engage with the process of technological innovation in any critical way, this too could have major implications for the rules governing the functioning of the basic interorganizational structure.

Finally, the pharmaceutical example suggests the line that might be taken in elaborating the perspective presented here. The next step must be through study of innovation processes in other areas having comparable properties, and innovation in pharmaceutical products is an obvious candidate. We would then look at the nature of integration between the pharmaceutical industry and the medical profession. Liebenau suggests that these relations took their modern form in the period around World War I (somewhat earlier than was the case for roentgenology). "Pharmaceutical suppliers moved from a position in which they had been the servants of physicians, supplying whatever materials the doctors might need, to one in which they were the leaders, guiding medics toward new products" (Liebenau 1986, 83). Are structural relations, mutual interdependencies, and innovation processes different from those disclosed in the present analysis? Is the innovation process itself evolving, as has been suggested to be the case in regard to diagnostic imaging? The road toward a comprehensive theory of technological innovation being charted here is a long and arduous one. In my view it is the one road that seems to offer the hope of a theory that does justice to the complexity, and the importance, of the phenomenon.

Notes

Chapter 1

1. The term *medical technology* is conventionally taken collectively to denote drugs, devices (instrumentation), and procedures. For a preliminary, but useful, discussion of how innovation processes in these three areas of medical technology might differ, see Gelijns (1989).

2. Economists often make the distinction between *technological change*, which means the development of new technology, and *technical change*, which refers to changing practice (that is, to the gradual replacement of one technology by another).

3. But one must not use these kinds of figures in any way to characterize the technology "as such." The costs of operating a given instrument depend not only on the intensity with which it is used, but on the whole cost structure of health care also. In addition much will depend on the number of years over which the capital cost is amortized in any calculation. Thus a senior British radiologist worked out the cost per brain scan in his hospital, as of 1977/78, at precisely £15, or approximately $30 in terms of the exchange rates of those days (Thomson 1979).

4. In his *Principes de Médicine Expérimentale*, begun in the 1860s, Bernard wrote, "La médicine expérimentale règne en plein aujourd'hui. C'est moi qui fonde la médicine expérimentale, dans son vrai sens scientifique; voilà ma prétention." (Experimental medicine reigns supreme today. It is I who have founded experimental medicine, in its true scientific sense: that is my claim." (my translation))

5. Perhaps the technology required further refinement. Wunderlich's thermometer was over a foot long. The pocket clinical thermometer was developed later by the English physician Clifford Allbutt.

6. The classic treatment of specialization in American medicine has been provided by Rosemary Stevens (see Stevens 1971).

7. The rise in public expenditure on health care is relevant in this context. In the United States, in 1929, total expenditure on health care was $3.6 billion, or $29 per capita. By 1950, these figures had risen to $12 billion and $78 per capita (Fain, Plant, and Milloy 1977, 466). This does not, however, correspond to an equivalent rise in the volume of services provided, since there is also evidence that the *costs* of health services rose faster than the general consumer price index in this period (Somers and Somers 1961, 193).

8. Warner offers the following comment on the book: "Although infidelity to his sources ultimately leads Brown's work to collapse, the publicity his thesis received brought the issue of the ideological uses of medical science to the forefront of medical historians' attention" (Warner 1985).

9. The General Electrical Company had been formed in 1892 through a merger of the Thomson-Houston Company and Edison General (an amalgamation of Thomas Edison's various manufacturing interests). The new firm had an income in the tens of millions.

10. There are suggestions that noncommercial considerations also played a role in strategic decision making in this area. Reich quotes sources suggesting that the Coolidge tube, though initially seen as too expensive to be competitive, was marketed because the firm felt obligated to make available a device so "useful to humanity" (Reich p. 89 and note 80). There is an obvious temptation to see the search for legitimacy to which Brown alludes in this episode.

11. Though the prior claims of photography as a diagnostic imaging technology could also be argued (see Reiser 1978, 57).

12. With the introduction of nuclear medicine (the diagnostic use of radiopharmaceuticals, stimulated by the development of the gamma camera in the early 1950s)—to a large degree under the control of radiologists—the fruits of a second innovation process could be harvested, although they had to be shared. With the introduction of radioactive isotopes into hospitals, used in both diagnosis and (radio) therapy, the need arose for control of the safety and operating effectiveness of equipment. These functions came to be located in departments of hospital physics, staffed by graduate—and sometimes research—physicists. As a result, hospital physicists occupy an important place in the control, and sometimes the use, of much of the hospital's advanced technology. Though rivalries and problems of demarcation may ensue, the physicists, no less than the radiologists, have an interest in technological advance.

Chapter 2

1. Kamien and Schwartz, who have been developing a theory of innovation based in decision theory, have discussed the various uncertainties that influence a firm's *choice* of the speed at which to develop. They have distinguished between technical uncertainty (how much will it cost to develop?); market uncertainty (what is the competition doing?) and uncertainty as to eventual profit (see Kamien and Schwartz 1982, chapter 4).

2. In the United States four firms (GE, Picker International, Philips, and Siemens) controlled more than 70 percent of the total diagnostic imaging market as of 1978, and nearly 90 percent of the market in conventional x-ray equipment (Hamilton 1982, 60).

3. In April 1987 two of the largest firms in the sector, GEC/Picker and the Medical Systems Division of Philips, announced an impending merger, with the increasing cost of development work given as the reason. Early in 1988 the merger was called off. The two firms had been unable to agree on their respective financial contributions.

4. Firms with experience with the medical technology market know this well enough, of course. Those without experience either assume it by virtue of general assumptions about U.S. compared with other (e.g., European) markets, or soon learn it from informants. This will be well illustrated by the CT scanner example in chapter 5.

5. This is of course only true within limits. People can choose whether a certain set of symptoms merit a visit to the doctor, or whether they will—for example—try to cure themselves with over-the-counter drugs. There are also alternative health care systems in many Asian countries and some of these, together with old European systems such as homeopathy, are seemingly of growing importance also in the West. But hospitals have a monopoly on most of the services they provide.

6. Like business historians such as Chandler, Williamson distinguishes between what he calls "hierarchy" (which refers to the internalization of functions by the organization) and "market" (which can be viewed as an alternative means of allocating risks, resources, and benefits) (Williamson 1975, 1981) and argues that under many circumstances "hierarchy works better than markets"—as some put it, the visible hand works better than the invisible. For a useful overview of this literature see Barney and Ouchi 1986.

7. It is interesting to note that Williamson's theory provides some rationale for von Hippel's finding regarding the user role in innovation: the "customer active paradigm." We can see that for some industries, possibly including those that provide health care, some partial integration of suppliers is rational. We could then see this steering influence on innovation as consonant with Granstrand's notion of "quasi-integration."

8. Though economists do recognize that there are conditions under which markets do not form.

9. This is a figure of speech. The apparatus at that time was large and unmovable; it had to be maintained as a fixed installation, although it could be connected to the various wards of the hospital by long electrical leads.

10. For a discussion of the differences between these two forms of explanatory accounting in historiography, see McCullagh 1984, 205 et seq.

11. "There exists a great chasm between those, one the one side, who relate everything to a single central vision . . . and on the other side , those who pursue many ends, often unrelated and even contradictory" (Berlin 1957, 7–8).

12. Granovetter (1985) has argued that sociological theorizing generally differs from economic theorizing in the sense that whereas the former typically employs an "oversocialized" conception of man, the latter typically employs an "undersocialized" one, quoting in the process James Duesenberry's quip that "economics is all about how people make choices; sociology is all about how they don't have any choices to make." This is a different point from the one I making here. Nevertheless, Granovetter's attempt to find an adequate middle way for the sociological understanding of economic action is useful, and I shall return to it later.

13. "By the time assessment results are obtained, the availability of a new improved embodiment of the technology makes the validity of the results suspect." (Hillman 1988)

14. Hård conceives of an "ideal type" as "a schematic representation of the basic characteristics inherent in an area of technics" : a heuristic, a sense of how, in abstract terms, a technology might be realizable. An "archetype" is one real working embodiment of this general image, rooted in particular historical circumstances, and which serves as a model for later "types" (Hård 1987).

15. Interview with R. Cowen, June 1982.

16. Radiologists define the accuracy of a diagnostic test in terms of two parameters, sensitivity and specificity. These may be based on comparison with the results of autopsy or biopsy (the "gold standard"), or with some other test taken as definitive. *Sensitivity* answers the question "If the disease is present, how likely is the patient to have a positive test?" while *specificity* answers the question "if the disease is absent, how likely is the patient to have a negative test?" (see Horwitz et al. 1984).

Chapter 3

1. It seems clear, from a number of early accounts, that Langevin was led to believe that ultrasound might have some therapeutic possibilities, rather similar to the therapeutic uses of x-rays. There is, however, no evidence that he worked on this.

2. The Radio Corporation of America was established in 1919 when American Marconi was taken over by General Electric and other U.S. industrial interests. Between 1919 and 1930 the firm, later RCA, was controlled by outside interests. One seat on the board was reserved for a representative of the U.S. Navy (Graham 1986, 35–38)

3. Hackmann points out that some effort was devoted to the examination of German equipment and the interviewing of top German sonar scientists. "The most comprehensive report was written by Batchelder of the Submarine Signal Company and a member of the U.S. Technical Mission in Europe" (p. 294).

4. The origin, of his work, Denier writes, is a study he chanced upon, "called 'Essai de l'Influence du Son sur l'Economie,' a thesis written in the faculty of medicine of Montpellier and presented on 7 February 1828. . . . Our colleague cites a number of facts recounted by Percy, surgeon to the Grande Armée: soldiers retrieving from the Rhine and the Vistula huge fishes which had been caused to perish by the endless artillery fire; deaf mutes who whilst not hearing the noise nevertheless sustained emotional shock from the shock of the sound. These facts, produced by acoustic phenomena, seemed to me curious, and it is in order to better understand them that I have thrown myself into the world of acoustics" (Denier 1946a) (my translation).

5. "Mais c'était alors un veritable poste de detection de radar qui ne serait pas sortir du domaine du laboratoire et n'aurait pu entrer dans la clinique" (Denier 1946a). ("But it would in effect be a radar detection post, which would not leave the domain of laboratory. It would not have been possible to introduce it in the clinic." (my translation)) Thus the connection with radar, which for others was to prove the crucial stimulus, was for Denier an indication of the impossibility of the technique in practical diagnostics.

6. Thus almost precisely the system that Denier a few years before had seen as never leaving the laboratory.

7. This is a later technique involving injection of an opaque medium into the carotid artery. Although it could show up intracranial blood vessels invisible in air studies, it often has to be given under anesthetic and could lead to headaches and other complications.

8. This remark excludes the therapeutic, and especially surgical, uses of ultrasound at high-energy levels which were also being investigated, notably by a group at the University of Illinois. Yoxen reports that Heuter and Ballantine moved into this area, before Heuter left to join Raytheon in 1956.

9. This device made use of the fact that the velocity of sound in water is some 2 × 10^5 less than that of radio waves in air, so that "by substituting an ultrasound beam for an aircraft's normal radar beam, it was possible for a pilot to practice 'flying' over a tank of water containing a small scale map of enemy territory" (Hill 1973).

10. Ellen Koch's recent dissertation provides a detailed account of the disputes with both NIH and his host institutions in which Wild became embroiled. Unfortunately this study, which offers a comparative analysis of main U.S.-based ultrasound programs, became available to me after this chapter was written. (See Koch 1990.)

11. This had been noted by McLoughlin and Guastavino in 1949, but was now being rediscovered.

12. The concept of *gain* refers to the technique whereby signals are amplified in proportion to the distance they have travelled. In this way less intense echoes received from deeper in the body are amplified more than more intense echoes, making them more accessible to analysis than they would otherwise be.

13. As we saw earlier, these workers had been making use of a transmission method, and their failure had been due to the masking by the skull of transmission patterns produced by the brain.

14. This is more apparently true in neurology, where a number of studies were stimulated by Leksell's findings. For example, the question of the source of the "midline echo," which was the basis of the diagnostic method, was taken up by the Dutch neurologist de Vlieger in Rotterdam (de Vlieger and Ridder 1959).

15. Donald has provided a number of anecdotal but also quite detailed accounts of the origins of his ultrasound work. What follows is largely based on Donald 1980.

16. Interview with Professor J. MacVicar, 27 July 1981

17. The period from 1957 to 1960 has been termed the period of "greatest radiological caution": the proportion of pregnant women x-rayed fell to a minimum.

18. Fetal skull measurement, or fetal cephalometry, was an important obstetrical application of radiology. Its use had been pioneered in the 1920s by, among others, H. Thoms. By the 1930s it was widely used. (See Hiddinga 199.)

19. One final anecdote rounds off these stories. By 1957, in Lund, the results obtained by Leksell and by Edler were a topic of lunchtime conversation. Donald's work being as yet unpublished, the suggestion that the possibilities of the technique in obstetrics merited investigation led to Bertil Sundén being encouraged to investigate a number of pregnancies using the Krautkrämer flaw detector. The patterns of echoes obtained from the pregnant uterus were too complex to be of

any clinical use. When Donald's paper appeared in 1958, and the Lund group learned of the two-dimensional device that had been built, Sundén was sent to Glasgow to spend three weeks working with Donald. Returning to Lund, he brought knowledge of ultrasound complementary to what had been developed there. (Interview with Dr. B. Sundén, 1 October 1981)

20. Some firms were nevertheless uninterested in the new market. Douglas Gordon, a radiologist with extensive interests as an inventor, has described how one long-established British firm making industrial flaw detection equipment resisted for many years his attempts (from 1964 on) to persuade them to enter the medical market. Subsequently Gordon himself made the minor modification to their instrument needed for ophthalmic work and marketed it through a firm of his own. That a one-inventor firm could do this is indicative of how low the barriers to entry were (Gordon, personal communication).

21. I owe this comment, and much of the information on Kelvin Hughes and Smiths, to an interview with Mr. Brian Fraser (22 April 1982). In the late 1960s Mr. Fraser was principally involved in the nondestructive testing side of ultrasonic technology with the firm. I am extremely grateful to Mr. Fraser for his help.

22. Information from interview with Dr. B. Sundén, University of Lund.

23. I owe the following account to a private communication from Professor Hellmuth Hertz, for whose help I am very grateful.

24. Information provided by Siemens (UK). It is interesting to note what happened to Krautkrämer. In the early 1970s, the firm was taken over by the American company Branson, which had itself been taken over a few years previously by the pharmaceutical company Smith Kline and French (SKF). SKF themselves, through their subsidiary Smith Kline Instrument Company (based in Philadelphia), had moved into echographic equipment much earlier. By the mid 1960s there was an A-scan instrument, the Ekoline 20, which was in use among neurologists and ophthalmologists, and there was a collaborative project looking to develop a B-scan instrument.

25. In a 1974 review Dr. Holmes implies as much. At a 1965 conference Holmes refers to his group having just acquired a Porta scanner (Holmes 1966). The Porta scanner seems to have been produced by Physionics Engineering, of Longmont, Colorado. This appears, from the illustrations, to be the same device as that designed by Wright. In his 1974 review Holmes thanks Picker for the photograph used, which suggests that Picker had taken over Physionics Engineering. I have been unable to confirm this, however.

26. See for example an editorial in the *American Journal of Roentgenology*, "The general radiologist and ultrasound," published in 1976 (127, 867).

27. In neurology the dispute was to be transformed shortly afterward by the arrival of the CT brain scanner. This effectively put to rest neurological interest in ultrasound for quite some time.

28. One estimate suggests that by 1972 eighty instruments were to be found in British wards alone. Oakley points out that the first reference to the technique in a standard British textbook of prenatal care (Browne's) appears in the 1970 edition, where it is noted, "The apparatus is costly and not yet available in many

clinics, but it undoubtedly has great value, especially in difficult cases" (Oakley 1984, 165).

29. Oakley quotes the following figures for the percentages of pregnant women receiving ultrasound scans at two leading London hospitals: Queen Charlotte's 1973 48 percent rising to 97 percent in 1978; The London Hospital 1973 26.5 percent rising to 61.5 percent in 1978. Quotations of the views of leading obstetricians are also to be found there. (see Oakley 1984, 165–167.)

30. Sonicaid started on a very small scale in 1962, making flaw detectors, and moved into the medical area in 1967 with a very low price fetal heart detector (costing only £95 or $300). In just five years, between 1968 and 1973, the firm's turnover increased from £20000 ($70000) to £950 000 ($3.2 million). In 1969 they decided to move into fetal monitoring, and started a development project that reached fruition in 1971. (All conversions have been made on the basis of purchasing power parity exchange rates obtaining in the relevant year.) Sonicaid nearly came to disaster in the late 1970s. Tom Brown, who had left Nuclear Enterprises, persuaded them to back him in developing a stereoscopic scanner. Although the instrument worked, the firm's founder recalls "We couldn't persuade the doctors of the desirability of viewing in three dimensions." The firm lost a great deal of money on this venture. I am very grateful to Mr. R. Cowen for this account (interview with R. Cowen, 1 June 1982).

31. Information provided by Mr. Brian Fraser.

32. This kind of question was arising in other areas both inside and outside medicine. It is central to the developments to be discussed in chapter 5.

33. A 1977 article in *Modern Healthcare* quotes the ultrasound product manager at Picker as having doubts regarding further specialization of instrumentation. Picker had developed a special transducer arm for imaging very small areas (such as the thyroid) but was unsure about eventual marketing (Trasker 1977).

34. Compare the example of magnetic disk storage discussed in chapter 2.

Chapter 4

1. Interestingly, Barnes had been a classmate of Hardy's at Harvard. Both of them, together with Perkin Elmer—founder of the instruments firm bearing his name— had been members of the Harvard physics class of 1923.

2. This is the same Smiths Industries that had taken over Kelvin Hughes not long before, and which we came across in chapter 3 in connection with the Donald/ Brown ultrasound work.

3. Interview with Mr. K. Lloyd Williams, 10 May 1982.

4. For reasons that will become obvious later, Lloyd Williams, who participated, was subsequently to describe this conference as "in some ways premature."

5. *Case reports* are publications detailing experiences with small numbers of patients, sometimes one. Although they are conventionally accepted as something other than research, they have a definite importance in medicine. See Fletcher, Fletcher, and Greganti 1982.

6. Interview with Dr. Francis Ring, 23 July 1986. I am particularly grateful to Dr. Ring for the time and trouble he took in explaining a good deal of the technical and social history of thermography to me.

7. Gershon Cohen continued to publish in the same style as before until his death in 1971. In 1969 he announced a new, inexpensive thermometric device which he called a mammometer, and in 1970 a further device called the Biocomp (or Bioelectronic Thermographic Computer). The extent of his commitment to breast screening by mammography—supplemented by these various techniques—was beginning to involve him in considerable controversy. An interesting obituary of Gershon Cohen is to be found in the *American Journal of Roentgenology* (Hermel 1971).

8. Although there is a logical progression in the work there is also a very great difference in style from that of Gershon Cohen, whose writing reflects a different view of how the medical profession ought to evaluate thermography. The contrast is clearly brought out if we compare Isard's paper with a piece written by Gershon Cohen in 1970, in *Cancer.*

Statistics are, of course, necessarily impressive, but they often fail to convey the importance of specific experiences which have been crucial to preserving individual lives. From a series of 35 consecutive breast cancers which have passed through our offices in the past two years, 4 have been selected as examples of women in whom prompt biopsy occurred chiefly because of positive thermographic and thermometric findings. Supportive signs by physical examination and mammography were lacking, yet in each instance biopsy established the presence of a Stage I cancer. (Gershon Cohen, Hermel, and Murdock 1970)

9. This distribution of veins can be studied by infrared *photography*. Indeed it had been in the late 1940s. Infrared photography is a quite different technique, which shines infrared radiation on a surface and looks at the reflection. Thermography, it will be recalled, records the infrared radiation emitted by the body being studied. Infrared photography of the breast had been studied following a suggestion of Massopust in 1948, to the effect that the superficial venous pattern (which is what the technique displays) was altered significantly by a carcinoma. The finding was not borne out by other studies, and interest in use of the technique faded in the early 1950s. It had, however, been demonstrated that the distribution of superficial veins in the breasts was frequently asymmetrical.

10. Texas Instruments (TI) had begun as a geophysical services company in the 1930s. In the early 1950s the firm had switched to the then new transistor market, and with considerable success. By the early 1960s TI was the world's largest supplier of semiconductors. At the end of 1962 the firm was given the original contract for development and preproduction work on the Minuteman II intercontinental ballistic missile (see Dorfman 1987, 175–179, 216).

11. In the editorial preface to a new journal, *Thermology*, launched in April 1985, reference is made to "the strategic publication of statistical studies based upon thermograms of such technical inferiority that a number of experienced thermologists, this writer among them, refused to participate in their interpretation. Even so, the studies were published, unaccompanied by the experimental

data. . . . There was no open review . . . no critical analysis . . . and as a result breast thermography suffered a setback" ("The Building of a Journal," *Thermology* 1: 1 (1985)).

12. This comes from an internal Barr & Stroud memorandum "Background to B&S involvement in medical thermal imaging," June 1986, kindly made available to me by Mr. D. C. Lorimer.

13. Interview with Dr. C. Jones, 19 November 1985.

14. An isothermal display is analogous to the contour lines on a map. It simply means that at selected intervals lines are drawn on the display linking points of equal temperature. Here, typically, differences of 0.5 degrees are used.

15. Lloyd Williams took up a senior surgical appointment at a nearby Bath hospital in 1964, partly in order to further his thermographic interests.

16. In the way that they were doing at that time in regard to EMI's development of the CT scanner (see chapter 5).

17. This account is based on a personal communication (3 November 1986) from Mr. D. C. Lorimer, Product Sales Manager with Barr & Stroud, for whose help I am very grateful.

18. Interview with Dr. H. J. Isard, July 1987.

19. Recall how in the case of ultrasound this arose in respect of the establishment of safety, as when Donald drew attention to his attempts at interesting anatomists in the effects of ultrasonic radiation.

Chapter 5

1. In their study of the origins of radio astronomy Edge and Mulkay note how, at the Cavendish Laboratory in the 1950s, radio astronomers and x-ray protein crystallographers shared similar problems. "Both had to await independent development in computer design before achieving their technical goals" (p. 132).

2. Ventriculography and encephalography both entail injecting air into the cavities of the brain in place of its normal fluid content. They date from the end of the First World War. Angiography is a later technique that could show up intracranial blood vessels invisible in air studies. Involving the injection of an opaque medium into the carotid artery, it frequently had to be given under anesthetic. Each of these techniques could lead to patients suffering from recurring headaches and other complications. As Jennett writes "The noninvasive alternatives of electroencephalography and later of echoencephalography . . . were disappointingly limited in what they could offer" (Jennett 1986, 58).

3. This section is based on the invaluably detailed account of the invention of the CT scanner given by Süsskind (1981).

4. IBM first marketed a commercial computer that used (2200) transistors in 1955 (Braun and Macdonald, 69).

5. Hounsfield was not the only one thinking along these lines. A group of physicists and radiologists from the University of Canterbury, New Zealand, were working along similar lines. By the time this group published their idea of a Fourier reconstruction analysis of x-ray data, and the results of applying it to a dog's shin

bone, it was already too late. Their paper "Computer aided transverse body-section radiography" appeared in the *British Journal of Radiology* after the first papers of Hounsfield and Ambrose—which went much further—had been submitted. (See Peters, Smith, and Gibson, 1973.)

6. The only experience the firm had had with medical technology was a short-lived involvement in infrared thermography. As we saw in chapter 4, this never got beyond the construction of a few prototypes.

7. Recall from chapter 4 that just a couple of years later DHSS would approach the Ministry of Defense Research Establishment at Aldermaston, asking them to design a thermograph for breast screening. Thermographs cost between £5 and £7,000 (around $20,000), CT scanners more than twenty times as much.

8. This discussion is based on this invaluable reference. The notion of *constraints* is an important one. During the process of successive iteration, negative values for certain pixels will be obtained on occasion. Knowing as we do that the x-ray density at a point cannot be negative, we can introduce as a constraint that all negative values are to be put equal to zero. Another constraint that was to become important in scanning was to make use of known x-ray densities, such as bone. Gordon et al point out that although "constrained ART" is less mathematically elegant than the unconstrained version, it produces more accurate reconstructions.

9. Larsen and Rogers, in their study of Silicon Valley, argue that skepticism regarding the value of patents is widespread in the electronics industry (Larsen and Rogers 1984). Hamilton notes that in the medical devices industry it is not usual; EMI's battery of CT patents was unusual (Hamilton 1982, 61).

10. Recall how theoretical considerations had also suggested the greater receptivity to innovation of a competitively organized hospital system such as that of the United States. See chapter 2.

11. See chapter 2.

12. Fourier had first shown that any complex transient waveform can be expressed as the sum of an infinite series of simple sine waves of suitable frequency, etc. The Fourier transform is a mathematical relationship that transforms a transient waveform with *time* as the variable to a set of components of variable *frequency*.

13. Information kindly provided by Dr. G. M. K. Hughes, Vice President, Systems and Communication, Pfizer Pharmaceuticals, New York.

14. This varied from one country to the other of course. In the Netherlands leveling off only set in from about 1981/82 (STG 1987, vol. I, 12). For the technology, and for manufacturers, the U.S. market was of dominant importance, being something like three times the size of that of the EEC as a whole (in terms of numbers installed in 1980).

15. Telephone conversation with Dr. Hughes, 12 September 1988.

16. In 1976, with only 6.3 percent of group sales it had provided 19.0 percent of profits. In 1977 11.0 percent of sales had generated 20.0 percent of group profits (Süsskind 1981, 66).

17. A ray of light, however, was the settlement regarding EMI's infringement of patent suit with Ohio Nuclear. Technicare, Ohio Nuclear's parent, was taken over

by Johnson & Johnson in October 1978. Johnson & Johnson agreed to resolve the litigation with a $15 million out-of-court retrospective royalty payment to EMI and to take out a worldwide license under EMI's patents.

Chapter 6

1. Interview with Professor Mallard, 3 June 1982. I am grateful to Professor Mallard for his careful explanation of MRI and of some of the particularities of the field of medical physics.

2. Electron Spin Resonance, or ESR, refers to the resonance phenomena associated with unpaired electrons, for example in ion radicals. The phenomenon was discovered by the Russian scientist Zavoyskii in 1944. The first commercial ESR spectrometers were produced by Varian Associates in the late 1950s.

3. Running costs of the helium-cooled superconducting magnet his scanner used were very high, and by late 1973 the NIH money was virtually finished. Applications for further funding to the NIH and to the American Cancer Society were turned down. Convinced that the NMR "establishment" was against him, Damadian set out to raise money through a charitable organization that he set up with his brother-in-law: the Citizens Campaign for New Approaches to Cancer (Kleinfield, 71–75).

4. Damadian and Lauterbur have different accounts of this since Damadian apparently claims that Lauterbur stole his idea, whereas Lauterbur apparently claims that he knew little of Damadian's work beyond its general application to biopsy samples. The relations between the two men come vividly out of Kleinfield's fascinating story.

5. These profiles had in fact been investigated in the early 1950s, but only twenty years later was their potential as a means of generating true two-dimensional images recognized.

6. Interestingly, it appears that he had been led publicly to speculate on these applications by the editor of *Nature*. The first version of his paper had been rejected, and the request was made that he refer to possible uses of his method. In revision he "tacked on" the above paragraph (Kleinfield 1985, 62). The fact that Lauterbur refers the NMR-tumor connection to Wiseman and not to Damadian himself, even though Wiseman followed Damadian's lead, was apparently one of the things that incensed the latter.

7. Interview with Professor E. R. Andrew, 12 February 1982.

8. By means of different imaging sequences it is possible to generate images more or less dependent upon the three basic NMR parameters: proton density, T_1, and T_2.

9. By this time these were not the only R&D programs based outside medicine. Not only was work beginning in a number of companies (see following), but both Lauterbur's group in Stony Brook and Smidt's group in Delft were (once more) active. The Nottingham work does adequately illustrate the *sorts* of adaptations made in the research programs.

10. Recall Howry's appeal to the quality of *his* first ultrasound images, compared with Wild's appeal to diagnostic statistics. (See chapter 3.)

11. Picker International was formed by GEC (not to be confused with GE), which merged its existing medical interests with those of the American company Picker, which it took over. GEC/Picker were themselves developing a resistive magnet system like that which they now agreed to support in Nottingham.

12. Technicare, the parent company of Ohio Nuclear, was taken over by Johnson & Johnson in 1978, from which point a major investment in magnetic resonance imaging was made. Technicare subsequently hired Waldo Hinshaw, who had been working with Andrew.

13. In order that the company's activities not attract the attention of GE and Johnson & Johnson—which knew the name FONAR—Damadian operated his company under the name RAANEX II Corporation for the first two years.

14. Interview with Professor C. Cramer, Coordinator R&D, Philips Medical Systems, 11 January 1983.

15. Industry has invested billions of dollars in development in anticipation of a rapidly growing and ultimately very large market. For hospitals, whether despite or because of the investment entailed (the FONAR scanner cost $1.6 million in 1985), an MRI scanner became one of the status symbols of the late 1980s. All this amounts to a considerable strengthening of the resources and the status of radiology—with both to some degree at the cost of other specialties or other approaches to medicine.

16. These high field strength systems are not at this time (Spring 1991) available as production models. I understand that there are a number of manufacturers willing and able to build one in response to a specific order. (Peter Morris, personal communication)

17. The major x-ray producers seem gradually to be assuming control of the market. Since this chapter was written, it has been brought to my attention that the Aberdeen-based firm M&D Technology Ltd. has ceased production. I am grateful to Peter Morris for this information.

Chapter 7

1. Pinch has discussed this notion of "black boxing" in terms of agreement over observing and accounting. When agreement is eventually reached as to the significance of a given observational report (knowledge claim), at the same time "the validity of the observational process used to produce the report is established." That is, the questions that could be asked, and in the past perhaps had been asked, about the observing instruments themselves, are now set aside. Pinch writes: "The observing process is now a 'black-boxed' instrument (along with, of course, all the associated instrumental practices). Examples of such black-boxed instruments are Geiger counters and chart recorders" (Pinch 1986).

2. One question given relatively little consideration was patient safety, and it was pointed out that little information was available on the x-ray dosage levels administered by the different models on the market (Kehoe 1976).

3. A senior physicist, reflecting on his experience of collaborating with physicians on an imager development program, remarked "it's been interesting. They have a different approach to problems, I'd say. . . . They are highly empirical. . . . They are willing to try things . . . and if it produces a result, a positive result, an effective result, they'll go on using it, even when there's no understanding of how it works."

4. Carney considers available strategies in relation to collectives formed from within a set ("ecology") of comparable organizations. The argument here focuses on the complementarity of the strategies with which manufacturers and their clients defend a shared interest.

5. This view is supported by the fact that manufacturers are apparently unwilling at this time to make high field strength MRI equipment, capable of producing this kind of image, commercially available. I am grateful to Peter Morris for this information.

References

Aarts, N. J. M. (1977)"Current status of thermography in Europe." *Acta Thermographica* 2:41.

Abernathy, W. J., and K. B. Clark (1985) "Innovation: mapping the winds of creative destruction." *Research Policy* 14:3.

Ackermann, R. (1981) In *The Technological Imperative in Medicine:* S. Wolfe and B. B. Berk eds. London and New York: Plenum p. 9.

Aitken, H. G. J. (1978) "Science, technology and economics: the invention of radio as a case study." In *The Dynamics of Science and Technology*, R. Krohn, E. Layton and P. Weingart eds. Dordrecht and Boston: Reidel.

Alfidi, R. J., MacIntyre, W.J. and Meaney, T.F. (1975) "Experimental studies to determine applications of CAT scanning to the human body," *American Journal of Roentgenology* 124:199.

Ambrose, J. (1975) "A brief review of the EMI scanner." *British Journal of Radiology* 49: 605

Ambrose, J. M., R. Gooding and A. E. Richardson (1975) "EMI scanning and intracranial tumours." *Brain* 98:569.

Ambrose, J. (1973) "Computed transverse axial scanning (tomography) Pt 2: clinical application." *British Journal of Radiology* 46:1023

American Hospital Association (1983) "Nuclear magnetic resonance: Guideline Report" *AHA Hospital Technology Series* vol. 2,8 Chicago, Ill.

Amsterdamska, O. forthcoming "The clinical connection: contexts of research on bacterial variation."

Amsterdamska, O., and L. Leydesdorff (1989) "Citations: indicators of significance?" *Scientometrics* 15:449.

Andrew, E. R., P. A. Bottomley, W. S. Hinshaw, G. N. Holland, W. S. Moore, and C. Simaroj (1977) "NMR images by the multiple sensitive point method: application to larger biological systems." *Physics in Medicine and Biology* 22:971.

Arrow, K. (1962)"Economic welfare and the allocation of resources for inventive activity." In *The Rate and Direction of Inventive Activity*, R. R. Nelson. Princeton, N. J: Princeton University Press.

Astheimer, R. W., and E. M. Wormser (1959)"Instrument for thermal photography." *Journal of the Optical Society of America* 49:184.

Atsumi, K. (1977) "The past studies and developments on bio-medical thermography in Japan." *Acta Thermographica* 2:67.

Baker, H. L. (1976) "Computed tomography and neuroradiology: a fortunate primary union." *American Journal of Roentgenology* 127:101

Baker, H. L., J. K. Campbell, O. W. Houser, and D. F. Reese (1975) "Early experience with the EMI scanner for study of the brain." *Radiology* 116:327.

Baldwin, W. L., and G. L. Childs (1969) "The fast second and rivalry in research and development." *Southern Economic Journal* 36:18.

Ballantine, H. T., R. H. Bolt, T. F. Heuter and G. D. Ludwig (1950) "On the detection of intracranial pathology by ultrasound." *Science* 112:525.

Ballantine, H. T., T. F. Heuter and R. H. Bolt (1954) "On the use of ultrasound for tumor detection." *Journal of the Acoustical Society of America* 26:581

Banta, H. D. (1984) "Embracing or rejecting innovations: clinical diffusion of health care technology." In *The Machine at the Bedside*, S. J. Reiser and M. Anbar eds. Cambridge: Cambridge University Press, 64–94.

Barclay, A. E. (1949) "The old order changes" *British Journal of Radiology* 22:300.

Barley, S. R. (1986) "Technology as an occasion for structuring: evidence from observations of CT scanners and the social order of radiology departments." *Administrative Sciences Quarterly* 31:78.

Barnes, R. B. (1963) "Thermography of the human body." *Science* 140: 870.

Barnes, R. B. (1964) "Thermography." *Annals of the New York Academy of Science* 121:34.

Barney, J. B., and W. G. Ouchi, eds. (1986) *Organizational Economics*. London and San Francisco: Jossey Bass.

Barron, S. L. (1950) "The development of the electrocardiograph in Great Britain." *British Medical Journal* 1:720.

Berk, R. N., and S. S. Siegelman (1988) "The value of early publications on efficacy of MR imaging." *American Journal of Roentgenology* 151:1240.

Berlin, I. (1957) *The Hedgehog and the Fox*. New York: Mentor.

Bijker, W. (1984) "Techniekgeschiedenis: een mogelijke basis voor theorieen over technologische ontwikkeling?" *Jaarboek voor Geschiedenis van Bedrijf en Techniek* 1:44.

Bijker, W. (1986) "The social construction of bakelite." In *The Social Construction of Technological Systems*, W. Bijker, T. P. Hughes and T. J. Pinch eds. Cambridge, Mass: The MIT Press.

Blume, S. S. (1981) 'Technology in medical diagnosis: aspects of its dynamic and impact." In *Innovation in Health Policy and Service Delivery*, C. Altenstetter ed. Cambridge Mass: Oegleschlager Gunn and Hain.

Blume, S. S. (1982) "Explanation and social policy: 'the' problem of social inequalities in health." *Journal of Social Policy* 11:7.

Bohnenkamp, H., and H. W. Ernst (1931) "Untersuchungen zu den Grundlagen des Energie- und Stoffwechsels: Uber die Strahlungsveluste des Mensch." *Archiev für der gesamtliche Physiologie* 228:41.

Branemark, P. I. (1967) "Biologic and clinical evaluation of infra red thermography," *Journal de Radiologie et d'Electrologie* 48:69.

Braun, E., and S. MacDonald (1982) *Revolution in Miniature.* Cambridge: Cambridge University Press.

Brecher, R. and E. Brecher (1969) *The Rays: A History of Radiology in the United States and Canada.* Baltimore: Williams and Wilkins.

Brown, E. R. (1979) *Rockefeller Medicine Men.* Berkeley: California University Press.

Burch, G. E. (1961) "Developments in clinical electrocardiography since Einthoven." *American Heart Journal* 61:324.

Burns, T. R. (1985) "The development of alternative energy technologies: entrepreneurs, new technology, and social change." Berlin: International Institute for Environment and Society, Discussion Paper 85-6.

Burrows, E. H. (1986) *Pioneers and Early Years: A History of British Radiology.* Alderney, C. I.:Colophon.

Callon, M. (1985) "Some elements of a sociology of translation: domestication of the scallops and fishermen of St. Brieuc Bay." In *Power Action and Belief,* J. Law ed. Sociological Review Monographs. London: Routledge and Kegan Paul.

Calman, N. (1984) "Women and medication: an empirical examination of the extent of womens' dependence on medical technology in the early detection of breast cancer." *Social Science & Medicine* 18:561.

Campbell, S., and G. B. Newman (1971) "Growth of the fetal biparietal diameter during normal pregnancy." *British Journal of Obstetrics and Gynaecology* 78:513.

Carlson, W. B. (1984) *Invention, Science and Business: the Professional Career of Elihu Thomson, 1870–1900* Ph.D. diss., University of Pennsylvania, Philadelphia.

Carmichael, J. H. E., and R. J. Berry (1976) "Diagnostic x-rays in late pregancy and in the neonate." *Lancet* 1:351.

Carney, M. G. (1987) "The strategy and structure of collective action." *Organization Studies* 8:341.

Cobet, R., and F. Bramigk (1924) "Uber messung der Warmtestrahlung des menschlichen Haut und ihre Klinische Bedeutung." *Deutsches Archiev für klinische Medizin* 144:45.

Cockburn, C. (1985) *Machinery of Dominance.* London: Pluto Press.

Collins, A. J., and J. A. Cosh (1970) "Temperature and biochemical studies of joint inflammation." *Annals of Rheumatic Diseases* 29:386.

Collins, A. J., E. F. J. Ring, J. A. Cosh, and P. A. Bacon (1974) "Quantitation of thermography in arthritis using multithermal analysis I: the thermographic index." *Annals of Rheumatic Diseases* 33:113.

Collins, H. M. (1974) "The TEA set: tacit knowledge and scientific networks." *Science Studies* 4:165.

Colton, R. M., ed. (1982) *Analyses of Five National Science Foundation Experiments to Stimulate Increased Technological Innovation in the Private Sector.* Section V (medical instrumentation) Washington DC: National Science Foundation.

Comroe, J. H., and R. D. Dripps (1981) "Scientific basis for the support of biomedical science." In *Biomedical Innovation*, E. B. Roberts, R. I. Levy, S. N. Finkelstein, J. Moskowitz, and E. J. Sondik, eds. Cambridge, Mass: The MIT Press.

Cooper, L. S., T. C. Chalmers, M. McCally, J. Berrier, and H. S. Sacks (1988) "The poor quality of early evaluations of magnetic resonance imaging." *Journal of the American Medical Association* 259:3277.

Cormack, A. M. (1963) "Representation of a function by its line integrals, with some radiological applications." *Journal Applied Physics* 34:2722.

Cosh, J. A. (1966) "Infra red detection in the assessment of rheumatoid arthritis." *Proceedings of the Royal Society of Medicine* 55:Suppl. 88–92.

Cosh, J. A., and F. Ring (1967) "Techniques of heat detection used in the assessment of rheumatic diseases." *Journal de Radiologie et d'Electrologie* 48.

Creditor, M. C., and J. B. Garrett (1977) "The information base for diffusion of technology: computed tomography scanning." *New England Journal of Medicine* 297:49.

Damadian, R., L. Minkoff, M. Goldsmith, M. Stanford and J. Koutcher (1976) "Field focusing nuclear magnetic resonance (FONAR): visualization of a tumor in a live animal" *Science* 194:1430.

Damadian, R., M. Goldsmith, and L. Minkoff (1977) "NMR in cancer XVI: FONAR image of the live human body." *Physiological Chemistry and Physics* 9:97.

Damadian, R., K. Zaner, D. Hor, T. DiMaio, L. Minkoff, and M. Goldsmith (1973) "Nuclear magnetic resonance as a new tool in cancer research: human tumors by NMR." *Annals of the New York Academy of Sciences* 222:1048.

Damadian, R. (1971) "Tumor detection by nuclear magentic resonance." *Science* 171:1151.

Davis, Aubrey B. (1981) "Life insurance and the physical examination: a chapter in the rise of American medical technology." *Bulletin of the History of Medicine* 55:392.

Délégation Générale à la Recherche Scientifique et Technique (DGRST). *Les Technologies Biomédicales au Japon*. Paris: La Documentation Française. (1980)

Denier, A. (1946a) "Les ultrasons. Leurs applications au diagnostic: ultra-sonoscopie et la thérapeutique: ultra-sonothérapie" *Journal de Radiologie et de l'Electrologie* 27:481.

Denier, A. (1946b) "Ultrasonoscopie. Note de M. André Denier" *Comptes Rendus de l'Academie des Sciences* 222:785.

Desjardins, A. U. (1929) "The low status of radiology in America." *Journal of the American Medical Association* 92.1035.

DiMaggio, P. J., and W. W. Powell (1983) "The iron cage revisited: institutional isomorphism and collective rationality in organizational fields." *American Sociological Review* 48:147.

Dodd, G. D., A. Zermeno, L. Marsh, D. Boyd, and J. D. Wallace (1969) "New developments in breast thermography." *Cancer* 24:1212.

Dodd, G. D. (1977) "Present status of thermography, ultrasound and mammography in breast cancer detection." *Cancer* 39:2796.

Dodd, G. D., J. D. Wallace, I. M. Freundlich, L. Marsh, and A. Zermeno (1969) "Thermography and cancer of the breast." *Cancer* 23:797.

Dognon, A., and L. Gougerot "Les possibilités de l'ultra-sonoscopie." *Paris Medical* 22: (February 1947) 81.

Donald, I. (1974) "New problems in sonar diagnosis in obstetrics and gynecology." *American Journal of Obstetrics and Gynecology* 118:299.

Donald, I. (1980) "Medical sonar: the first 25 years." In *Recent Advances in Ultrasound Diagnosis* 2, A Kurjak, ed. Amsterdam: Excerpta Medica 4.

Donald, I., J. MacVicar, and T. G. Brown (1958) "Investigation of abdominal masses by pulsed ultrasound." *Lancet* 1:1188.

Dorfman, N. S. (1987) *Innovation and Market Structure—Lessons from the Computer and Semiconductor Industries*. Cambridge, Mass: Ballinger.

Dosi, G. (1982) "Technological paradigms and technological trajectories: a suggested interpretation of the determinants of technological change." *Research Policy* 11:147.

Draper, J. W., and C. H. Jones (1969) "Thermal patterns of the female breast." *British Journal of Radiology* 42:401.

Draper, J. W., and J. W. Boag (1971) "The calculation of skin temperature distributions in thermography." *Physics in Medicine and Biology* 16:201.

Dussik, K., F. Dussik, and L. Wyt (1947) "Auf dem Wege zur Hyperphonographie des Gehirnes." *Wiener Medizinische Wochenschrift* 38–9:425.

Eden, M. (1984) "The engineering-industrial accord:inventing the technology of health care." In *The Machine at the Bedside*, S. J. Reiser and M. Anbar, eds. Cambridge: Cambridge University Press, 49.

Edge, D. O., and M. J. Mulkay (1976) *Astronomy Transformed*. London and New York: Wiley.

Edler, I. (1955) "The diagnostic use of ultrasound in heart disease." *Acta Medica Scandinavica Supplement* 308:32.

Ehrenreich, B., and D. English (1973) *Complaints and Disorders: the Sexual Politics of Sickness*. New York: The Feminist Press.

Evans, K. T., and J. H. Gravelle (1973) *Mammography Thermography and Ultrasonography in Breast Disease*. London: Butterworth.

Evens, R. G. (1980) "Nuclear magnetic resonance: another new frontier for radiology?" *Radiology* 136:795.

Evens, R. G., and R. G. Jost (1973) "Economic analysis of computed tomography units." *American Journal of Roentgenology* 127:191.

Fain, T. G., K. C. Plant, and R. Millay, eds. (1977) *National Health Insurance*. Public Documents Series. New York and London: R. R. Bowker Co.

Figlio, K. (1977) "The historiography of scientific medicine." *Comparative Studies in Society and History* 19:262

Fletcher, S., R. Fletcher, and M. Greganti (1981) "Clinical research: trends in general medical journals 1946-1976." In *Biomedical Innovation*, E. Roberts et al., eds. Cambridge Mass: The MIT Press.

Freeman, C. (1977) "The economics of research and development" in I. Spiegel-Rösing and D. de S. Price, eds. *Science Technology and Society*, London and Beverly Hills: Sage.

Freeman, H., F. E. Linder, and R.F. Nickerson (1937) "Bilateral symmetry of skin temperature." *Journal of Nutrition* 13:39.

French, L. A., J. J. Wild, and D. Neal (1951) "The experimental application of ultrasonics to the localization of brain tumors." *Journal of Neurosurgery* 8:198.

Freundlich, I. M. (1980) "Medical aspects of thermography." In *Imaging for Medicine*, S. Nudelman and D. B. Patton, eds. New York and London: Plenum Press, 400.

Friedman, P. J. (1988) "The early evaluations of MR imaging." *American Journal of Roentgenology* 151:860.

Fries, S. D. (1988) "2001 to 1994: political environment and the design of NASA's space station system." *Technology and Culture* 29:568.

Fujimura, J. H. (1987) "Constructing 'do-able' problems in cancer research: articulating alignment." *Social Studies of Science* 17:257.

Galison, P., (1988) "History, philosophy, and the central metaphor." *Science in Context* 2:197.

Garroway, A. N., P. K. Grannell, and P. Mansfield (1974) "Image formation in NMR by a selective irradiation process." *Journal of Physics (C): Solid State Physics* 7:1457.

Gautherie, M. (1983) "Thermobiologic assessment of benign and malignant breast diseases." *American Journal of Obstetrics and Gynecology* 147:861.

Gawler, J., J. W. D. Bull, G. H. DuBoulay, and J. Marshall (1974) "Computer assisted tomography (EMI Scanner): its place in investigation of suspected intracranial tumours" *Lancet* 1:419.

Geison, G. L. (1979) "Divided we stand: physiologists and clinicians in the American context." In *The Therapeutic Revolution*, M. Vogel and C. Rosenberg, eds. Philadelphia: Pennsylania University Press.

Gelijns, A. C. (1989) *Technological Innovation: comparing development of drugs, devices and procedures in medicine*. Washington DC: National Academy Press.

Gershon Cohen, J. (1967) "Medical thermography." *Scientific American* 216:94.

Gershon Cohen, J., S. M. Berger, J. D. Haberman, and E. E. Brueschke (1964) "Advances in thermography and mammography." *Annals of the New York Academy of Sciences* 125:283.

Gershon Cohen, J., M. B. Hermel, and M. G. Murdock (1970) "Priorities in breast cancer detection." *New England Journal of Medicine* 283:82.

Gibbons, M. and R. Johnson (1974) "The roles of science in technological innovation." *Research Policy* 3:220.

Glasser, D. (1934) *Wilhelm Conrad Roentgen and the Early History of Roentgen Rays* Springfield, Ill: Ch C Thomas.

Gordon, R., G. T. Herman, and S. A. Johnson (1975) "Image reconstruction from projections" *Scientific American* 233: (October) 56.

Graham, M. B. W. (1986) *RCA and the VideoDisc: the Business of Research*. Cambridge: Cambridge University Press.

Granstrand, O. (1979) *Technology, Management and Markets.* Gothenburg: Chalmers University of Technology.

Grannell, P. K., and P. Mansfield (1975) "Microscopy in vivo by nuclear magnetic resonance." *Physics in Medicine and Biology* 20:477.

Granovetter, M. C. (1973) "The strength of weak ties" *American Journal of Sociology* 78:1360.

Granovetter, M. C. (1982) "The strength of weak ties: a network theory revisited." In *Social Structure and Network Analysis,* P. V. Marsden and N. Lin, eds. London and Beverley Hills: Sage.

Granovetter, M.C. (1985) "Economic action and social structure: the problem of embededdness." *American Journal of Sociology* 91:481.

Greer, A. C. (1984) "Medical technology and professional dominance theory," *Social Science and Medicine* 18:809.

Gros, Ch., C. Vrousos, P. Bourjat, and J. L. Mugel (1968) "Place actuel de la thermographie dans l'exploration des tumeurs abdominales." *Annales de Radiologie* 11:773.

Güttner, W., G. Fiedler, and J. Pätzold (1952) "Uber ultraschallabbildungen am Menschlichen Schädel." *Acustica* 2:148.

Hackmann, W. D. (1984) *Seek and Strike: Sonar, Antisubmarine Warfare and the Royal Navy 1914–1954.* London, HMSO.

Hackmann, W. D. (1986) "Sonar research and naval warfare 1914–1954: a case study of a twentieth century science." *Historical Studies in the Physical and Biological Sciences* 16:83.

Hamilton, B. (1982) *Medical Diagnostic Imaging Systems.* New York: Frost & Sullivan.

Hard, M. (1984) "Ideal type-archetype-type: the typology of nineteenth century refrigeration technology." (Paper prepared for a Workshop on Social and Historical Studies of technology, Enschede, the Netherlands).

Hardy, J. D. (1934) "The radiation of heat from the human body: I An instrument for measuring the radiation and surface temperature of the skin," *Journal of Clinical Investigation* 13:593.

Hardy, J. D., and C. Muschenheim (1934) "The radiation of heat from the human body: IV The emission, reflection and transmission of infra-red radiation by the human skin." *Journal of Clinical Investigation* 13:817.

Harris, D. L., W. P. Greening, and P. M. Aichroth (1966) "Infra red diagnosis of a lump in the breast." *British Journal of Cancer* 20:710.

Hartley, K., and J. Hutton (n.d.) "The UK medical equipment industry: structure, performance and public policy." University of York, ISER discussion paper #99.

Harvey, A. McG. (1981) *Science at the Bedside: Clinical Research in American Medicine 1905–1945.* Baltimore: Johns Hopkins University Press.

Hermel, H. B. "Obituary of Jacob Gershon Cohen." *American Journal of Roentgenology* 112:417.

Herrick, J. F., and F. H. Krusen (1954) "Ultrasound and medicine. A survey of experimental studies" *Journal of the Acoustical Society of America* 26:236.

Hertz, C. H. "The interaction of physicians, physicists and industry in the development of echocardiography." *Ultrasound in Medicine and Biology* 1 (1973) 3

Hiddinga, A. (1987) "Obstetrical research in the Netherlands in the nineteenth century." *Medical History* 31:281.

Hiddinga, A. (1991) "X-ray technology in obstetrics: measuring pelves at the Yale School of Medicine." In *Medical Innovation in Historical Perspective*, J. V. Pickstone, ed. London: Macmillan.

Hiddinga, A. and S. S. Blume (Forthcoming) "Technology science and obstetric practice: the origins and transformation of cephalo-pelvimetry." *Science Technology and Human Values.*

Hill, C. R. (1973) "Medical ultrasonics: an historical review." *British Journal of Radiology* 46:899.

Hillman, B. J. (1988) "The assessment of MR imaging." *American Journal of Rontgenology* 151:858.

Hindle, Brooke (1981) *Emulation and Invention*. New York and London: W. W. Norton.

Hinshaw, W. S., P. A. Bottomley, and G. N. Holland (1977) "Radiographic thin section image of the human wrist by nuclear magnetic resonance." *Nature* 270:722.

Hinshaw, W. S., E. R. Andrew, P. A. Bottomley, G. N. Holland, W. S. Moore, and B. S. Worthington (1978) "Display of cross sectional anatomy by nuclear magnetic resonance imaging." *British Journal of Radiology* 51:273.

Hinshaw, W. S. (1974) "Spin mapping: the application of moving gradients to NMR." *Physics Letters* A48:87.

Hinshaw, W. S. (1976) "Image formation by NMR: the sensitive point method." *Journal of Applied Physics* 47.

Hippel, E. von (1976) "The dominant role of users in the scientific instrument innovation process." *Research Policy* 6:212.

Hippel, E. von, and S. N. Finkelstein (1979) "Analysis of innovation in automated clinical analyzers." *Science and Public Policy* 6.

Hodges, P. C. (1945) "The development of diagnostic x-ray apparatus during the first 50 years." *Radiology* 45:438.

Holmes, F. L. (1984) "Lavoisier and Krebs: the individual scientist in the near and deeper past." *Isis* 75:131.

Holmes, J. H. and D. Howry (1963) "Ultrasonic diagnosis of abdominal diseases." *American Journal of Digestive Diseases* 8:12.

Holmes, J. H., W. Wright, E. P. Meyer, G. J. Posakony, and D. H. Howry (1965) "Ultrasonic contact scanner for diagnostic applications." *American Journal of Medical Electronics* 4:147.

Holmes, J. H. (1980) "Diagnostic ultrasound during the early years of A. I. U. M." *Journal of Clinical Ultrasound* 8:299.

Holmes, J. H. (1974) "Diagnostic ultrasound: historical perspective" In *Diagnostic Ultrasound*, D. L. King, ed. St Louis: C V Mosby, 1.

Holmes, J. H. (1966) In *Diagnostic Ultrasound*, Grossman et al eds. Proceedings of the First International Conference, Pittsburgh, 1965. New York: Plenum.

Horwitz R. I., A. R. Feinstein, W. B. Credé, and J. D. Clemens (1984) "Does technology work? judging the validity of clinical evidence." In *The Machine at the Bedside*, S. J. Reiser and M. Anbar, eds. Cambridge: Cambridge University Press, 193.

Hounsfield, G. N. (1973) "Computerized transverse axial scanning (tomography): description of system." *British Journal of Radiology* 46:1016.

Howell, J. D. (1984) "Early perceptions of the electrocardiogram: from arrythmia to infarction." *Bulletin of the History of Medicine* 58:83.

Howell, J. D. (1986) "Early use of x-ray machines and electrocardiographs at the Pennsylvania Hospital 1897 through 1927." *JAMA* 255:2320.

Howry, D. H. (1957) "Techniques used in ultrasonic visualization of soft tissues." In *Ultrasound in Biology and Medicine*. E Kelly, ed. Washington DC: American Institute of Biological Science, 49.

Howry, D. H., and W. R. Bliss (1952) "Ultrasonic visualization of soft tissue structures of the body." *Journal of Laboratory and Clinical Medicine* 40:579.

Howry, D. H., D. A. Stott, and W. R. Bliss (1954) "The ultrasonic visualization of carcinoma of the breast and other soft tissue structures." *Cancer* 7:354.

Hughes, T. (1983) *Networks of Power: Electrification in Western Society 1880-1930.* Baltimore: Johns Hopkins University Press.

Hutchison, J. M. S., M. A. Foster, and J. R. Mallard (1971) "Description of anomalous E. S. R. signals from normal rabbit liver." *Physics in Medicine and Biology* 16:655.

Hutchison, J. M. S., J. R. Mallard, and C. C. Goll (1974) "In vivo imaging of body structures using proton resonance." In *Proceedings of the 18th Ampere Congress*, P. S. Allen et al eds. Nottingham: University of Nottingham, 283.

Hutton, J. (n. d.) "The role of the government in the research, development and commercialisation of CT scanning and NMR in the UK." York: IRISS, University of York.

Hutton, J. and K. Hartley (1985) "The influence of Health Service procurement policy on research and development in the UK medical capital equipment industry." *Research Policy* 14:171.

Iglehart, J. K. (1977) "The cost and regulation of medical technology: future policy directions." *Milbank Memorial Fund Quarterly* 55:25.

Isard, H. J., B. J. Ostrum, and R. Shilo (1969) "Thermography in breast carcinoma." *Surgery Gynecology and Obstetrics* 128:1289.

Isard, H. J., W. Becker, R. Shilo, and B. J. Ostrum (1972) "Breast thermography after four years and 10,000 studies." *American Journal of Roentgenology* 115:811.

Janus, C. L,. and S. S. Janus (1981) "Ultrasound: state of the art." *Journal of Clinical Ultrasound* 9:217.

Jellins, J., G. Kossoff, T. S. Reeve, and B. H. Barraclough (1973/75) "Ultrasonic visualization of breast disease." *Ultrasound in Medicine and Biology* 1:393.

Jennett, B. (1986) *High Technology Medicine: Benefits and Burdens.* Oxford and New York: Oxford University Press.

Jones, C. H., and J. W. Draper (1970) "A comparison of infrared photography and thermography in the detection of mammary carcinoma." *British Journal of Radiology* 43:507.

Jones, C. H., W. P. Greening, J. B. Davey, J. A. McKinna, and V. J. Greeves (1975) "Thermography of the female breast: a five year study in relation to the detection and prognosis of cancer." *British Journal of Radiology* 48:532.

Joskow, P. L. (1981) *Controlling Hospital Costs: the Role of Government Regulation.* Cambridge, Mass:The MIT Press.

Kamien, M. I., and N. L. Schwartz (1982) *Market Structure and Innovation.* Cambridge: Cambridge University Press.

Kehoe, Louise (1976) "The battle for the X-ray scanner market." *New Scientist.* 25: (November) 457.

Kelly, E., ed. (1957) *Ultrasound in Biology and Medicine.* Washington DC: American Institute of Biological Science.

Kessler, D. A., S. M. Pape, and D. N. Sundwall (1987) "The Federal regulation of medical devices." *New England Journal of Medicine* 317:357.

Klarman, M. E. (1974) "Application of cost-benefit analysis to the health services and the special case of technological innovation." *International Journal of Health Services* 4:325.

Kleinfield, S. (1985) *A Machine Called Indomitable.* New York: Times Books.

Knaus, W. A., S.A. Schroeder, and D.O. Davis (1977) "Impact of new technology: the CT scanner." *Medical Care* 15:533.

Knorr-Cetina, K. (1981) *The Manufacture of Knowledge.* Oxford: Pergamon Press.

Knorr-Cetina, K. (1982) "Scientific communities or transepistemic arenas of research? A critique of quasi-economic models of science." *Social Studies of Science* 12:101.

Koch, Ellen (1990) *American Research on Ultrasound, 1947-1962.* Ph.D. diss., University of Pennsylvania, Philadelphia.

Kossoff, G., E. K. Fry, and J. Jellins (1973) "Average velocity of ultrasound in the human female breast." *Journal of the Acoustical Society of America* 53:1730.

Kossoff, G. (1978) "Diagnostic ultrasound: the view from down under." *Journal of Clinical Ultrasound* 6:144.

Kossoff, G. (1975) "An historical review of ultrasonic investigations at the National Acoustic Laboratories." *Journal of Clinical Ultrasound* 3:39.

Kossoff, G. (1972) "Improved techniques in ultrasonic cross sectional echography." *Ultrasonics* 221.

Kossoff, G., and D. E. L. Wilcken (1967) "The CAL ultrasonic cardioscope." *Medical and Biological Engineering* 5:25.

Koutcher, J. A., M. Goldsmith, and R. Damadian (1978) "NMR in cancer X: a malignancy index to discriminate normal and cancerous tissue." *Cancer* 41:174.

Laitanen, H. A., and G. W. Ewing (1977) *A History of Analytical Chemistry*. Washington DC: American Chemical Society.

Larkin, G. (1983) *Occupational Monopoly and Modern Medicine*. London: Tavistock Press.

Larsen, J. K., and E. M. Rogers (1984) *Silicon Valley Fever*. New York: Basic Books.

Lasagna, L. (1972) "Special subjects in human experimentation." In *Experimentation with Human Subjects*, P. A. Freund, ed. London: Allen & Unwin, 262.

Latour, B. (1983) "Give me a laboratory and I will raise the world." In *Science Observed*, K. D. Knorr-Cetina and M. J. Mulkay, eds. London and Beverly Hills: Sage Books.

Latour, B. (1984) *Les Microbes, Guerre et Paix*. Paris: A. M. Métailié.

Latour, B. and S. Woolgar (1979) *Laboratory Life*. London and Beverly Hills: Sage Books.

Lauterbur, P. (1973) "Image formation by induced local interactions: examples of employing nuclear magnetic resonance." *Nature* 242:190.

Lauterbur, P. (1980) "Progress in N. M. R. zeugmatographic imaging." *Philosophical Transactions of the Royal Society of London* B289:483.

Law, J. and M. Callon "The life and death of an aircraft: a network analysis of technical change." (Paper prepared for a Workshop on Social and Historical Studies of Technology, Enschede, the Netherlands)

Lawrence, C. (1985) "Incommunicable knowledge: science, technology and the clinical art in Britain 1850-1914." *Journal of Contemporary History* 20:503.

Lawson, R. N. (1956) "Implications of surface temperatures in the diagnosis of breast disease." *Canadian Medical Association Journal* 75:309.

Lawson, R. N. (1957) "Thermography: a new tool in the investigation of breast lesions." *Canadian Medical Services Journal* 13:517.

Lawson, R. N. (1964) "Early applications of thermography." *Annals of the New York Academy of Sciences* 121:31.

Ledley, R. S. (1974) "Innovation and creativeness in scientific research: my experience in developing computerized axial tomography." *Computers in Biology and Medicine* 4:133.

Ledley, R. S., G. DiChiro, A.J. Luessenhop, and M.L. Twigg (1974) "Computerized transaxial x-ray tomography of the human body." *Science* 186:207.

Ledley, R. S., J. B. Wilson, and G. DiChiro (1974) "The ACTA scanner: the whole body computerized transaxial tomograph." *Computers in Biology and Medicine* 4:145.

Leibowitz, J. (1970) *The History of Coronary Heart Disease*. London: Wellcome Institute for the History of Medicine.

Leksell, L. (1956) "Echo-encephalography: I Detection of intracranial complications following head injury." *Acta Chirurgica Scandinavica* 110:301.

Lewis, T. (1911) "The electrocardiographic method and its relationship to clinical medicine." *Proceedings of the Royal Society of Medicine*.

Liebenau, J. M. (1986) *Medical Science and Medical Industry*. London: Macmillan.

Ling, C. R., M. A. Foster, and J. R. Mallard (1979) "Changes in NMR relaxation times of adjacent muscle after implantation of malignant and normal tissues." *British Journal of Cancer* 40:898.

Lloyd Williams, K., F. Lloyd Williams, and R. S. Handley (1960) "Infra red thermometry in clinical practice." *Lancet* 2:958.

Lloyd Williams, K., F. Lloyd Williams, and R. S. Handley (1961) "Infra red diagnosis of breast disease." *Lancet* 2:1378.

Lloyd Williams, K. (1964b) "Pictorial heat scanning." *Physics in Medicine and Biology* 9:433.

Lloyd Williams, K. (1964a) "Infra red thermography as a tool in medical research." *Annals of the New York Academy of Sciences* 121:99.

McCullagh, C. B. (1984) *Justifying Historical Descriptions*. Cambridge: Cambridge University Press.

MacKenzie, D. and J. Wajcman, eds. (1985) *The Social Shaping of Technology*. Milton Keynes: The Open University Press.

McKinlay, J. B. (1981) "From 'promising report' to 'standard procedure': seven stages in the career of a medical innovation." *Millbank Memorial Fund Quarterly* 59:374.

McLoughlin, R. P., and G. N. Guastavino (1949) "LUPAM: Localizador ultrasonoscopico para aplicaciones médicas." *Revista de la A. M. A.,* 421.

Mallard, J. R. "The quest for NMR imaging: a rather personal walk down Memory Lane." (MSS).

Mallard, J. R. (1981) "The noes have it! Do they? Silvanus Thompson Memorial Lecture." *British Journal of Radiology* 54: 531.

Mallard, J. R., J. M. S. Hutchison, W. A. Edelstein, C. R. Ling, M. A. Foster, and G. Johnson (1980) "In vivo n. m. r. imaging in medicine: the Aberdeen approach, both physical and biological." *Philosophical Transactions of the Royal Society of London* B289:519.

Mansfield, E. (1985) "How rapidly does new industrial technology leak out?" *The Journal of Industrial Economics* 34:217.

Mansfield, P., and P. K. Grannell (1973) *Journal of Physics (C);Solid State Physics* 6:L422.

Mansfield, P., A. A. Maudsley, and T. Baines (1976) "Fast scan proton density imaging by NMR." *Journal of Physics (E): Scientific Instruments* 9:271.

Mansfield, P., and A. A. Maudsley (1977) "Medical imaging by NMR." *British Journal of Radiology* 50:188.

Mansfield, P., I. L. Pykett, P. G. Morris, and R. E. Coupland (1978) "Human whole body line scan imaging by NMR." *British Journal of Radiology* 51:921.

Marx, Jean (1978) "NMR research: analysis of living cells and organs." *Science* 202:958.

Marx, L. (1979) *The Machine in the Garden: Technology and the Pastoral Ideal in America*. New York and Oxford: Oxford University Press.

Maulitz, R. (1979) "Physician versus bacteriologist: the ideology of science in clinical medicine." In *The Therapeutic Revolution,* M. Vogel and C. Rosenberg, eds. Philadelphia: Pennsylvania University Press.

Merton, R. K. (1973) *The Sociology of Science.* Chicago: Chicago University Press.

Miles, R. E., and C. S. Snow (1978) *Organizational Strategy, Structure and Process.* New York: McGraw-Hill.

Minkoff, L. (1985) "On the history of NMR scanning in medicine." *Physiological Chemistry and Physics* 17:1.

Moore, W. S., and G. N. Holland (1980) "Experimental considerations in implementing a whole body multiple sensitive point nuclear magnetic resonance imaging system." *Philosophical Transactions of the Royal Society of London* 289:429.

Mowery, D., and N. Rosenberg (1979) "The influence of market demand upon innovation: a critical review of some recent empirical studies." *Research Policy* 8:102.

Myers, G. (1985) "Texts as knowledge claims: the social construction of two biology articles." *Social Studies of Science* 15:593.

Myers, L. P., and S. A. Schroeder (1982) "Physician use of services for the hospital patient: a review." In *Issues in Hospital Administration,* J. B. McKinlay, ed. Cambridge, Mass: The MIT Press.

National Science Board (1982) *University-industry research relationships.* Washington DC: National Science Foundation.

Navarro, V. (1986) *Crisis, Health and Medicine: A Social Critique.* London: Tavistock Press.

Nelson, R., and S. Winter (1977) "In search of useful theory of innovation." *Research Policy* 6:36.

Neutra, R. R., S. E. Feinberg, S. Greenland, and E. A. Friedman (1978) "Effects of fetal monitoring on neonatal death rates." *New England Journal of Medicine* 299.

Oakley, A. (1984) *The Captured Womb.* Oxford: Blackwell.

Office of Technology Assessment (OTA) (1978) *Policy Implications of the Computed Tomography (CT) Scanner.* Washington DC: Congress of the United States.

Office of Technology Assessment (OTA) (1983) *The Emergence of NMR Imaging Technology: A Clinical, Industrial and Policy Analysis.* Washington D.C.: Congress of the United States.

Oldendorf, W. H. (1961) "Isolated flying spot detection of radiodensity discontinuities displaying the internal structure of a complex object." *I R E Transactions on Biomedical Electronics* 8:68.

Organization for Economic Co-operation and Development (1984) *Industry and University: New Forms of Co-operation and Communication.* Paris: OECD.

Pasveer, B. (1990) "Knowledge of shadows." *Sociology of Health and Illness* 11:360.

Paxton, R., and J. Ambrose (1974) "The EMI scanner: a brief review of the first 650 patients." *British Journal of Radiology* 47:530.

Peters, T. M., P. R. Smith, and R. D. Gibson (1973) "Computer aided transverse body-section radiography." *British Journal of Radiology* 46:314.

Phillips, A. (1979) "Organizational factors in R&D and technological change: market failure considerations." In *Research Development and Technological Innovation*, D. Sahal ed. D. C. Heath and Co.

Pinch, T. J. (1985) "Towards an analysis of scientific observation: the externality and evidential significance of observational reports in physics." *Social Studies of Science* 15:3.

Pinch, T. J., and W. Bijker (1984) "The social construction of facts and artefacts: or how the sociology of science and the sociology of technology might benefit each other." *Social Studies of Science* 14:399.

Posner, E. (1970) "Reception of Roentgen's discovery in Britain and the USA." *British Medical Journal* 4:357.

Potchen, E.J. (1981) "The value of efficacy studies." In *Biomedical Innovation* E.B. Roberts, R.I. Levy, S.N. Finkelstein, J. Moskowitz, and E.J. Sondit, eds. Cambridge Mass. and London: The MIT Press.

Rabkin, Y. M. (1987) "Technological innovation in science: the adoption of infrared spectroscopy by chemists." *Isis* 78:31.

Rapp, F. (1981) *Analytic Philosophy of Technology*. Dordrecht and Boston: Reidel.

Redisch, M. A. (1982) "Physician involvement in medical decision making." In *Issues in Hospital Administration*, J. B. McKinlay ed. Cambridge, Mass: The MIT Press.

Reese, D. F., P. C. O'Brien, G. W. Beeler, et al (1975) "An investigation for extracting more information from computerized tomography scans." *American Journal of Roentgenology* 124:177.

Reich, L. S. (1985) *The Making of American Industrial Research: Science and Business at GE and Bell, 1876–1926*. Cambridge: Cambridge University Press.

Reiser, S. J. (1978) *Medicine and the Reign of Technology*. Cambridge: Cambridge University Press.

Ring, E. F. J., and J. A. Cosh (1968) "Skin temperature measurement by radiometry." *British Medical Journal* (16 November) 448.

Ring, E. F. J., A. J. Collins, P. A. Bacon, and J. A. Cosh (1974) "Quantitation of thermography in arthritis using multi-isothermal analysis:II." *Annals of Rheumatic Diseases* 33:353.

Ring, F. (1977) "Quantitative thermography in arthritis using the AGA integrator." *Acta Thermographica* 2:172.

Rogers, E. M. (1983) *The Diffusion of Innovations*, 3rd ed. New York: Free Press.

Rosenberg, C. E. (1981) "Inward vision and outward glance: the shaping of the American hospital 1880—1914." In *Social History and Social Policy*, D. J. Rothman and S. Wheeler eds. New York: Academic Press.

Rosenberg, N. (1976) *Perspectives in Technology*. Cambridge: Cambridge University Press.

Rosenberg, N. (1982) *Inside the Black Box*. Cambridge: Cambridge University Press.

Russell, L. (1979) *Technology in Hospital*. Washington DC: Brookings Institution.

Saxton, H. M. (1973) "Seventy six years of British radiology." *British Journal of Radiology* 46:872.

Schaffer, S. (1988) "Astronomers mark time: discipline and the personal equation." *Science in Context* 2:115.

Schmookler, J. (1966) *Invention and Economic Growth.* Cambridge, Mass: Harvard University Press.

Shuman, W. P. (1988) "The poor quality of early evaluations of MR imaging: a reply." *American Journal of Roentgenology* 151:857.

Schumpeter, J. (1961) *Capitalism Socialism and Democracy.* London: George Allen and Unwin.

Sheedy, P. F. D.H. Stephens, and R.R. Hattery (1976) "Computed tomography of the body: initial clinical trials with the EMI prototype." *American Journal of Roentgenology* 127:23.

Sherwood, T. (1979) "Resources and decisions in clinical radiology" *Epidemiology and Community Health* 33:59.

Shils, E. (1956) *The Torment of Secrecy.* London: Heinemann.

Shrum, W. (1985) *Organized Technology: Networks and Innovation in Technical Systems.* Lafayette: Purdue University Press.

Sochurek, Howard (1987) "Medicine's new vision." *National Geographic Magazine* 171 (January).

Somers, M. M., and A. R. Somers (1961) *Doctors Patients and Health Insurance.* Washington, DC: The Brookings Institution.

Spiro, H. (1981) In *The Technological Imperative in Medicine,* S. Wolf and B. B. Berle eds. London and New York: Plenum Press

Stanley, R. J., S. S. Sagel, and P. F. Sheedy (1976) "Computed tomography of the body: early trends in application and accuracy of the method." *American Journal of Roentgenology* 127:53.

Starr, P. (1982) *The Social Transformation of American Medicine.* New York: Basic Books.

Stevens, R. (1971) *American Medicine and the Public Interest.* New Haven: Yale University Press.

Stewart, A., and G. W. Kneale (1968) "Changes in the cancer risk associated with obstetric radiology" *Lancet* 1:104.

Stewart, H. D. et al (1985) "Biological effects and ultrasound exposure levels." *Journal of Clinical Ultrasound* 13.

STG (Steering Committee on Future Health Scenarios) (1987) *Anticipating and Assessing Health Care Technology.* Dordrecht: Nijhoff.

Stocking, B. (1985) *Initiative and Inertia: Case Studies in the NHS.* London: Nuffield Provincial Hospitals Trust.

Stocking, B., and S. L. Morrison (1978) *The Image and the Reality.* Oxford: Oxford University Press.

Sundén, B. (1964) "On the Diagnostic Value of Ultrasound in Obstetrics and Gynaecology." *Acta Obstetrica et Gynaecologica Scandinavica* 43 (Supplmt 6).

Süsskind, C. (1981) "The invention of computed tomography." In *History of Technology*, vol.6, A. R. Hall and N. Smith eds. London: Mansell Publishing Co.

Temin, P. (1980) *Taking Your Medicine: Drug Regulation in the United States*. Cambridge Mass: Harvard University Press.

Thomson, J. (1979) "Cost effectiveness of an EMI brain scanner: an updated review." *Health Trends* 11:46.

Townsend, P., and N. Davidson (1982) *Inequalities in Health*. London: Penguin Books.

Trasker, M. R. (1977) "Ultrasound: the future survey device." *Modern Healthcare*. (December 24).

Tunnicliffe, E. J. (1973) "The British x-ray industry: a brief historical survey." *British Journal of Radiology* 46:861.

van den Daele, W., W. Krohn, and P. Weingart (1977) "The political direction of scientific development." In *The Social Production of Scientific Knowledge*, E. Mendelsohn, P. Weingart, and R. D. Whitley eds. Dordrecht and Boston: Reidel.

Vlieger, M. de, and H. J. Ridder (1959) "Use of echoencephalography." *Neurology* 1:216.

Wagner, J. L., and M. Zubkoff (1982) "Medical technology and hospital costs." In *Issues in Hospital Administration*, J. B. McKinlay ed. Cambridge, Mass: The MIT Press.

Walsh, D. (1897) *The Roentgen Ray in Medical Work*. London.

Warner, J. H. (1985) "Science in the historiography of American medicine." *Osiris* 2nd series, 37.

White, L. Jr., (1962) *Mediaeval Technology and Social Change*. Oxford: Oxford University Press.

Whitley, R. D. (1984) *The Intellectual and Social Organisation of the Sciences*. Oxford: Oxford University Press.

Wild, J. J., and D. Neal (1951) "Use of high frequency ultrasonic waves for detecting changes of texture in living tissue." *Lancet* 1:655.

Wild, J. J., and J. M. Reid (1952a) "Application of echoranging techniques to determination of structures of biological tissues." *Science* 115:226.

Wild, J. J., and J. M. Reid (1952b) "Echographic studies on tumors of the breast." *American Journal of Pathology* 28:839.

Wild, J. J., and J. M. Reid (1953) "The effect of biological tissues on 15Mc/s pulsed ultrasound." *Journal of the Acoustical Society of America* 25:270.

Williamson, O. E. (1975) *Markets and Hierarchies*. New York: The Free Press.

Williamson, O. E. (1981) "The economics of organization: the transaction cost approach." *American Journal of Sociology* 87:548.

Willocks, J., I. Donald, T. C. Duggan, and N. Day (1964) "Foetal cephalometry by ultrasound." *Journal of Obstetrics and Gynaecology of the British Commonwealth* 71:11.

Wise, M. Norton, (1988) "Mediating machines." *Science in Context* 2:77.

Wiseman, I. D., L. H. Bennett, L. R. Maxwell, M. W. Woods, and D. Burk (1972) "Recognition of cancer in vivo by nuclear magnetic resonance." *Science* 178:1288.

Young, I. R., D. R. Bailes, and M. Burl (1982) "Initial clinical evaluation of a whole body nuclear magnetic resonance (NMR) tomograph." *Journal of Computer Assisted Tomography* 6:1.

Yoxen, E. (1987) "Seeing by sound: a study of the development of medical images." In *The Social Construction of Technological Systems*, W Bijker, T Hughes, and T Pinch eds. Cambridge, Mass.: The MIT Press, 281.

Ziskin, M. C., M. Negin, C. Piner, and M. C. Lapayowker (1975) "Computer diagnosis of breast thermograms." *Radiology* 115:341.

Zola, I. (1975)."Medicine as an institution of social control." In *A Sociology of Medical Practice*, C. Cox and A. Mead eds. London: Collier Macmillan.

Index